SMART
TECHNOLOGIES

SMART TECHNOLOGIES

K. Worden, W. A. Bullough & J. Haywood

University of Sheffield, UK

World Scientific
New Jersey • London • Singapore • Hong Kong

Published by

World Scientific Publishing Co. Pte. Ltd.

5 Toh Tuck Link, Singapore 596224

USA office: Suite 202, 1060 Main Street, River Edge, NJ 07661

UK office: 57 Shelton Street, Covent Garden, London WC2H 9HE

British Library Cataloguing-in-Publication Data
A catalogue record for this book is available from the British Library.

SMART TECHNOLOGIES

ISBN 981-02-4776-1

Printed in Singapore by World Scientific Printers (S) Pte Ltd

Preface

The editors have been involved in a number of conferences covering the field of smart technologies. Two of them played particularly significant roles in the creation of this book. The first was a one-day meeting within the Institute of Physics Congress in Brighton during March 1998. This was formed from several keynote speeches by experts in different aspects of the field. Since then we, along with the other members of the Dynamics Research Group at the University of Sheffield, were proud to be able to host the International Conference on Smart Technology Demonstrators and Devices, which was held in Edinburgh during December 2001. This conference aimed to aid the transfer of smart technologies from the laboratory to the marketplace by focussing on working demonstrators and devices. In many ways it was a natural progression from the IOP meeting on theoretical basics to the Edinburgh meeting on fully-formed technology demonstrators.

Using the relationships built during these and other similar meetings, the editors have been able to invite some of the leading specialists in the various disciplines involved to help them put together a book that provides an up-to-date introduction to the rapidly developing world of smart technologies. Each author has been given a chapter in which to describe their particular enabling technology and it's relevance to the field. Because the authorship is taken from several different disciplines, the book hopefully reflects the multicultural nature of the field and will allow the reader to appreciate the different points of view that must be considered when designing smart technologies.

Particular emphasis has been placed on the use of examples of actual structures, materials, devices, systems and machines that have been developed with the concept of smart technology in mind. This hopefully highlights the broad range of applications that can benefit from the concept.

Anyone who needs to be briefed on the current status of these interdisciplinary technologies, or is interested in future developments in these fields will hopefully find this book helpful. It is presented in understandable and non-mathematical terms, making it accessible to engineers and scientists from an undergraduate level upwards.

Keith Worden, Bill Bullough and Jonathan Haywood

October 2002

Contents

Chapter 1

The Smart Approach — An Introduction to Smart Technologies

Keith Worden, William A. Bullough and Jonathan Haywood

Dynamics Research Group,
Department of Mechanical Engineering,
University of Sheffield,
Mappin Street, Sheffield S1 3JD, UK.

1.1 What Constitutes a Smart Technology?

The words *intelligent* and *smart* are often used as tools to market new products but sometimes this is done with little thought as to what this should mean. Some of these products may incorporate the highest of high technology but do they really possess an awareness of their situation? And are they then capable of reacting to it? These are the key attributes that a technology must exhibit for it to be considered a truly *smart technology*.

By situation we could mean the technology's environment, it's condition, or it's motion for example. The subsequent reaction could be to protect itself in some manner, instigate a repair to itself, or to adapt it's function so that it is tailored specifically to the situation.

Technologies with the ability to sense changes in their circumstances and execute measures to enhance their functionality under the new circumstances offer enormous benefits in performance, efficiency, operating costs and endurance.

1.2 Application of Smart Technologies

The breadth and depth of applications that could potentially gain an advantage from this concept, whether in medicine, engineering or physics, is vast. Each application will have many of it's own unique design criteria – determined by the intended function – but the creation of all smart technologies necessitates a solution to the same dilemma: how to integrate the fundamental abilities of awareness and reaction into a coherent system with the minimum of complexity and cost. This requires a design philosophy that is inherently interdisciplinary and requires knowledge of a number of different enabling technologies.

1.2.1 *An Interdisciplinary Field*

Awareness can be integrated into a system through the use of efficient sensing and measurement systems and the subsequent reaction may depend upon the application of innovative actuation technologies. It must be appreciated though, that the progression from awareness to reaction is not always straight forward. An ability to react is useless without the ability to know precisely when to react, and if more than one type of reaction is possible, then the ability to choose the appropriate reaction is also essential. In many cases sophisticated signal processing techniques and control strategies are therefore required.

The following chapters bring together many of these enabling technologies and hopefully illustrate their significance within smart technologies. The topics discussed are as follows:

Chapter 2 Sensing Systems Measurement systems for smart structures present particular challenges. Reliable strain, temperature, stress, etc. maps must be produced with the minimum of complexity. They should also readily interface into data reduction and processing systems in order to ascertain the detailed integrity of the structure under test.

The basic sensing and measurement specifications for structural monitoring are described in this chapter and examples of technologies which can be exploited in this context are given. Particular emphasis is applied to whole field techniques, which will enable distributed or quasi-distributed measurements including, in particular, optical fibre measurements.

Chapter 3 Vibration Control The increasing need for structures to be lighter, particularly in aerospace applications, generally makes them more susceptible to excessive vibration response. A number of satellites have been lost because of the vibration caused by thermal loads on the structure. One approach to this problem is to embed actuators and sensors, often piezoelectric, into the structure.

The general principles involved in the vibration control of smart structures are outlined in this chapter before a number of examples using this technology are described, including experiments on a smart antenna. Some thoughts on the future potential of smart structures are set down, and some of the outstanding problems are highlighted.

Chapter 4 Data Fusion Signal processing and computation are crucial elements in the implementation and operation of smart structures and systems. Sensors can provide an abundance of information about a system but it is useless without a means to reduce it to a manageable amount, identify the truly relevant information and then interpret it.

After an initial discussion as to how signal processing is incorporated into smart systems, this chapter presents two case studies, in the field of structural health monitoring, for illustration.

Chapter 5 Shape Memory Alloys A shape memory alloy (SMA) is an alloy that exhibits either large yet recoverable strains at a constant temperature (superelasticity), or apparently permanent strains that can be fully recovered on a change in temperature (thermal shape memory). The discovery of Nitinol kick-started interest in SMAs in the 1960s but the recent development of smart technologies has found many new applications for materials with this unusual property. Most of these applications involve SMAs being used as actuators, through their ability to convert thermal energy into motion and work output, in smart structures and machines.

The properties of shape memory alloys and the ways in which they can be employed in smart structures are outlined in this chapter. Examples of the use of SMAs in areas such as structural shape control and active damage control are also given.

Chapter 6 Piezoelectric Materials Naturally occurring piezoelectric materials (e.g. quartz, rochelle salt) have been known for over a hundred years and their use in the construction of force sensors employing the direct piezoelectric effect has been very successful. These generally offer the benefits of high stiffness and an associated high frequency bandwidth of measurement. With the development of ceramic 'artificial' piezoelectric materials, application of the inverse piezoelectric effect to produce actuators has become practical. These can provide large forces and ultra rapid actuation, but output displacements are extremely small.

This chapter explains the basic piezoelectric phenomenon and describes the different forms of piezoelectric actuators that have been developed to date. Methods for increasing their output displacement to enable them to be used in normal engineering tolerance products and devices are then considered. Application examples are given to illustrate the wide range of innovative uses which are now being found for piezoelectric materials.

Chapter 7 Magnetostriction Since the discovery of the exceptional low temperature magnetoelastic effects in the rare earth elements terbium (Tb), samarium (Sm) and dysprosium (Dy), renewed interest has been shown in magnetostrictive materials. The new 'giant' magnetostrictive compounds known by the acronym Terfenol offer a great potential for a wide variety of applications.

This chapter covers the development of the giant magnetostrictive materials for different applications and their system aspects, particularly in terms of control and power requirements for use in smart structures. Actual developed systems, including ones in active vibration control and sonar applications, are presented for illustration. Finally, a section on the use of active thin film technology based on magnetostrictive materials is discussed and depicted.

Chapter 8 Smart Fluid Machines Smart machines offer flexible operation through the ability to change their motion at the command of an electric signal and without a change in the hardware geometry. Analogous to the electric function generator which refers to various high capacity voltage sources to generate a signal shape, by governance of the time duration of exposure, the smart machine must deliver substantial forces, velocities and (most importantly) displacements; this makes very high acceleration implicit if multi-configuration output is to be had. The acceleration and displacement requirement with heat dissipation distinguish the smart machine from the structure — it will only carry light payloads.

The smart machine can be produced in various forms. It could be a change pattern (without stopping) carpet loom, a variable force shock absorber or a multi-shaped material handling device. Within limits dictated by inertial strain, overheating and driving stress saturation level, the required function can be accomplished by electromagnetic, electro-hydraulic, piezo latched motion and other means. This chapter touches on latest developments in this area and particularly on those involving electrorheological fluid and magnetorheological fluid hydraulic based techniques.

Chapter 9 Smart Biomaterials Following the implantation of a material or medical device, a number of possible outcomes may be realised. Unless particular care is taken in the selection of the material and the fabrication of the device, some of these can be disastrous. The bodies defences are highly adapted to identifying "foreign bodies" and dealing with these appropriately. This is fine if the foreign body is simple bacterium, which can be engulfed and destroyed, but not so good for the patient if the body is a "lump" of material purposely implanted to help restore some body function.

The bodies response to a material depends upon site of implantation and material. This chapter considers some of the typical responses, these include foreign body reaction, fibrous encapsulation, blood clotting, etc., and how materials are being engineered to circumvent these. It describes recent advances in materials engineering for tissue engineering, where the material actually aids the healing process by encouraging the growth of new tissue and the integration of the material/device. This is particularly important in the treatment of burns, or regeneration of nerves.

Chapter 10 Natural Engineering The natural world provides a vast number of ready-worked solutions to problems — it's the biologist's job to find out what those problems were. Once the question to the answer has been unravelled, together with the optimisations involved, the basic ideas can be applied elsewhere.

For smart systems the implementation of biological paradigms requires a shift away from traditional engineering. Examples are presented in this chapter that show applications at the material and structural level. Plants are presented as model systems for simple sensing and actuation.

Chapter 2

Sensing Systems for Smart Structures

Brian Culshaw

Department of Electronic and Electrical Engineering,
University of Strathclyde,
204 George Street, Glasgow G1 1XW, UK.

2.1 Introduction

The sensing system is the network of nerves within the smart structure. Its role is to monitor the state of the structure — to determine either loading conditions or the impact of physical deterioration on the structure's performance. Ideally it would be capable of detecting any changes in any part of the structural artefact. However even the biological model to which many of the smart structures advocates aspire focuses its sensory capabilities in specific areas determined through many generations of evolution (Fig. 2.1). The smart structure designer must conduct himself similarly so that concentrating the sensor array in the more vulnerable sections of the structure is an essential part of the sensor system design process. Some combination of intuition and system modelling can set this process in motion but, like nature, evolution must play an important part and learning from one generation to the next of designed artefact probably produces the most significant improvements.

This short overview will concentrate on sensor technology more than sensor system design. At present this is realistic. The system design process focuses on determining sensor location subject to the highest physical strain, to the largest temperature excursions, to the greatest exposure to

Fig. 2.1   The homunculus: a representation of sensor evolution in humans – the larger
the relative size of a particular area, the more sensitive.

the vagaries of the elements, or to the greatest vulnerability to chemical attack. As the whole design process becomes more integrated the procedures
for incorporating the nervous system will become embedded into the me-
chanical/structural monitoring procedure. Presently available design tools
are however some considerable way from this ideal.

2.2   Sensor Requirements in Smart Systems

Complete characterisation of the system using a sensor network is fundamentally unobtainable. At the very basic level, noise always inhibits system
performance so there are errors in the measurement process. These can be
corrected using some combination of system modelling and structural measurement based upon Kalman filters and similar procedures. The errors can
also be minimised through multiple measurement and integration processes.
They are however always there. There is also a significant issue concerning
the sheer volume of data and the ability to interpret measurements. The
volume of data would ideally meet the criteria of both spatial and tem-

poral sampling within the structure. A totally satisfactory measurement will need to meet requirements based on the number of nodes needed in a satisfactory finite element model of the structure's mechanical performance. This will rapidly become enormous especially when combining the spatial sampling with the necessary temporal sampling. Under these criteria, a structure such as a small aircraft would fill several CD-ROMs with data every second. Even if this could be acquired, analysing it down to a digestible answer would prove intractable. There would obviously also be considerable weight and power consumption penalties associated with a sensor system capable of performing at this level.

We must then focus the sensor system on what is required at the bare minimum level to make decisions. The first stage is to ascertain what these decisions may be. Essentially they fall into two categories. The first level in a structural system is "safe:unsafe", needs "further investigation to ascertain it is safe measured against whatever safety criteria are appropriate" and these obviously vary from application to application. This is the first phase in the sensing process. The second phase is "what are the loading conditions on this system and need the system respond to these loading conditions?". This of course fundamentally assumes that the necessary actuation response systems are in place. The definition of the "loading" may also be quite complex. The physical load is the obvious one. A thermal load — a change in temperature — may also induce a consequential physical load and will certainly introduce change in mechanical properties. Finally of course there can be chemical loads — changes in the pH of the local environment, chemical spillages, salt water spray, etc. — and these could also be responded to using appropriate reaction processes (Fig. 2.2).

The technical needs for monitoring physical variations are relatively straightforward to define. For a typical engineering structure measurements of strain to the order of 10μstrains and temperature to the order of a fraction of a degree centigrade are more than adequate over ranges which typically encompass the operational dynamic of strains to a few thousand microstrain and temperatures from -50 to $+150$°C. This of course excludes extreme environments such as gas turbine engines and down hole monitoring in oil exploration. It is interesting to reflect that the accuracy of the strain measurement and the accuracy of the temperature measurement requirement are roughly equivalent. Most structural materials have temperature coefficients of linear expansion in the region of 10 parts per million per °C so the two are obviously linked. There are most certainly exceptions

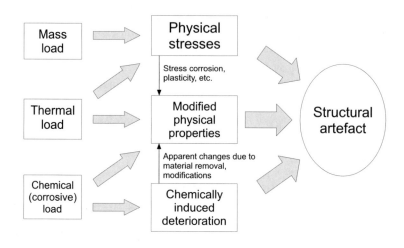

Fig. 2.2   Generalised structural loading.

to these broad guidelines so that individual specification must be drawn up for each application but experience has shown that the vast majority sit within an order of magnitude either way from these basic estimates. There is of course another aspect of measurement — how long should it be available and here the variations in aspirations are enormous. Some structures — missiles being the prime example — operate for but a few minutes. Others hopefully will last for generations and in these circumstances specifications requiring stability over 40 years and more are necessary but, for obvious reasons, have yet to be evaluated.

These basic parameters will apply equally to criteria for deterioration and loading though the latter must obviously be distinguished reliably from the former. Tests for structural deterioration should then be conducted under zero load conditions — which usually includes also an implicit zero temperature load. That is the measurements should always be made at the same temperature or at the very least compensated for temperature variations under zero mechanical load. Similarly loading measurements must include both mechanical load and thermal effects. The need for the ubiquitous supporting computer rapidly becomes obvious.

There has been quite appreciable success in assembling systems to measure mechanical parameters though data prediction and data interpretation remain as topics for further endeavour — as indeed do the sensor technolo-

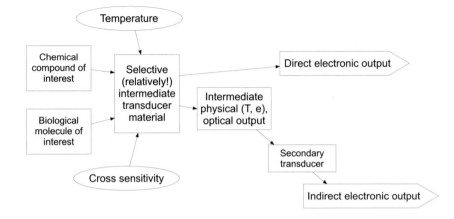

Fig. 2.3 Physical, chemical and biological sensing – the basic processes.

gies themselves and their successful incorporation into structures. The situation is however considerably different in the chemical domain. Chemical measurements inevitably imply chemical reaction and most chemical reactions are not reversible so that chemical measurements are often relatively unstable over long periods of time (Fig. 2.3). Biological systems cope with this through continual regeneration and the sensor designer has yet to invent the necessary equivalents. Additionally most chemical phenomena are not selective so that spurious responses are frequent. Similarly reagents often become contaminated ("poisoned") in use and must be replaced. Chemical reaction rates and processes are also very sensitive to temperature so thermal corrections are inevitably necessary. Consequently much of the discussion which follows will focus on physical rather than chemical measurements.

2.3 Sensor Technologies for Smart Systems

2.3.1 *The Options*

Sensing is a very specific art. The need is for an unambiguous electrical signal to furnish an input into a data analysis system connected to the structure of interest. The methods available for deriving this signal are many but all require some form of interface — our measurement is invari-

ably of a physical or chemical variable so that the electrical signal clearly
has to go through some form of interface process. Indeed there are many
systems — especially for chemical measurement — which involve two or
maybe three interfaces before the electrical signal eventually arrives. Small
wonder then — given all these interfaces — that stability and repeatability are recurrent issues in sensor technology. Many of these interfaces can
be unstable in time, frequently due to chemical changes and/or with temperature. I have often heard it said by those in the sensor industry that
primarily all sensor technologies are thermometers and the designer's first
task is to remove the temperature sensitivity.

These are generic sensor issues. In the smart structures context there
are more parameters to add. The need for multiple sensor inputs of different variables from a variety of locations within the structure of interest
and concomitant with that the requirement for reliable signal transmission
to the central collection area and preferably low or zero power electrical
consumption at the sensor itself. There are other needs including the most
important of all, that is for a consistent reliable intimate contact between
the sensor element and the parameter of interest (Fig. 2.4).

The options available include:

- Conventional transducers from the catalogue interfaced to increasingly powerful computing resources.
- Fibre optic sensor systems which can offer attractive possibilities
  in smart systems design.
- MEMS — both silicon and hybrid — which offer miniaturised precision and immense economies of scale in production volumes.
- Piezoelectric systems, particularly flexible polymeric large area sensors capable of multipoint addressing using techniques derived from
  flat screen interconnect concepts.
- Active surface films usually observed optically either visually or
  through camera systems.

There are no doubt others but these represent the major contributory concepts at the present time. Of these options fibre optic sensors and conventional approaches have been the most explored.

We look at each of these sensor options very briefly in turn. However,
we should always remember that the low cost computer system is central to
the whole discussion. Powerful readily available data analysis algorithms,
straightforward interfacing protocols, data correction and data filtering sys-

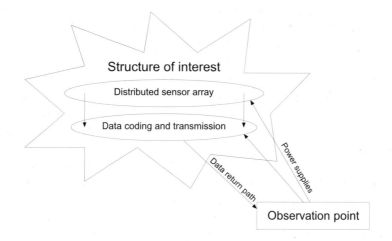

Fig. 2.4   Sensing requirements for smart structures.

tem and a plethora of other tools are an integral and essential contribution to the art of sensing and smart structures.

2.3.2   *Using Conventional Sensors*

Much has been achieved with conventional sensor technology — thermocouples, LVDTs, inclinometers, bonded strain gauges, vibrating wire gauges and many of the other established sensor systems (Fig. 2.5). Most — perhaps all — of the success of these systems have been in the physical domain. Chemical transducers with the necessary long term stability have really not become established. However most of these conventional transducers are relatively large and therefore intrude upon the aesthetics and/or the physical functionality of the structure of interest. Most of this points towards physically large structures as being the application area and in fact the majority of the structural assessment work using conventional sensors — apart from laboratory based testing — has been in civil engineering. Here not only can relatively large sensor elements be incorporated but there is also the physical space and structural strength to incorporate the necessary cable trays, power supplies, localised computer systems and all the other trimmings inherent in the smart system.

An excellent example of a structure extensively instrumented with conventional sensors is the Kingston Bridge in Glasgow (Fig. 2.6). Well over a thousand point sensors for strain, displacement, temperature and inclination are distributed throughout the bridge and its approaches. The aim of this sensor system was to monitor the bridge's behaviour in the build up to and during a major reconstruction to compensate for a serious slippage in the foundations of one of the main supporting piers. This culminated in the very delicate operation of lifting a bridge span weighing approximately 50,000 tons completely away from its original piers and replacing the bearings. The data collection system now monitors the bridge behaving normally.

This particular application typified the use of smart systems in civil engineering. During the build up to the final bearing replacement the sensor system provided data for a finite element model to ensure margins of safety in the bridge operation. Indeed at one stage the monitoring system did indicate potential difficulties resulting in a ten lane motorway bridge being closed for a few hours for careful inspection. The sensor system then ensured that the complex refurbishment and rebuilding programme was undertaken safely and without jeopardising the bridge's integrity. Now

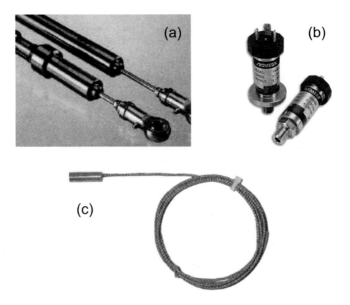

Fig. 2.5 Some conventional sensors: (a) LVDT displacement transducers; (b) pressure transmitters; (c) thermocouple.

Fig. 2.6   The Kingston Bridge, Glasgow, is extensively measured with over 1000 monitoring points distributed over 4km.

it simply produces reassurance that all is well. In the civil engineering context, smart systems are almost invariably there to provide a technical insurance rather than as an integral feature of system operation, with the very few exceptions in applications such as sway control in buildings and earthquake damage limitation.

2.3.3   *New Technologies — Fibre Optic Sensors*

Optical fibre sensors have provided some of the most intriguing prospects for the realisation of smart systems. Fibre sensors add a new dimension to structural integrity monitoring. In particular they enable full integration of the structure and the sensor — they can be embedded rather than attached so that intimate contact is assured. Additionally they can be realised in simple network architectures which are inaccessible using other sensing technology. A suitably treated single optical fibre can monitor at dozens, perhaps hundreds, of points along its length thereby removing the need for complex electronic multiplexing interconnect (Fig. 2.7).

Of the many approaches which exploit fibre optics in a sensing context smart systems fibre Bragg gratings are probably the most studied. The basic idea of a fibre Bragg grating (FBG) is that a periodic structure is written along the propagational direction of a single mode optical fibre. This

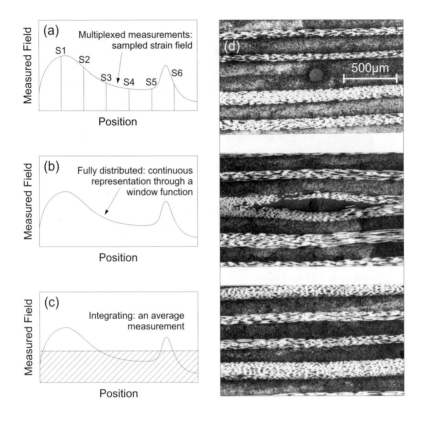

Fig. 2.7   Fibre optic sensing architectures: (a) multiplexed sensor arrays – available in all sensing technologies, including fibre optics; (b) distributed sensing — only readily available using fibre optic techniques; (c) integrating measurements, particularly useful in monitoring inhomogeneous materials, and easily implemented in fibre optics; (d) fibre optics embedded in carbon reinforced composites — the fibre is compatible both chemically and physically with the CRP.

periodic structure, which could perhaps extend over 10,000 wavelengths, has a period which is designed to produce constructive interference in the reflection direction at a specific wavelength and — for 10,000 periods — this reflection will have a fractional bandwidth of approximately 0.01% (Fig. 2.8). At a wavelength of 1.5µm, this corresponds to a bandwidth of around 0.15nm. Since this reflection grating is written in the fibre its period will change with temperature and strain so the reflected wavelength will change. Measuring the wavelength gives in principle a very simple technique for determining the period of the grating. Whether the change in the period is due to temperature or strain effects is something which must be determined. For static strain measurements this usually involves the presence of a temperature reference grating mounted free from the strain field and physically close to the measurement point. For dynamic strain, provided that the period of the strain variation is less than the thermal time

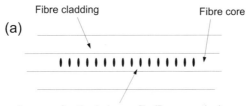

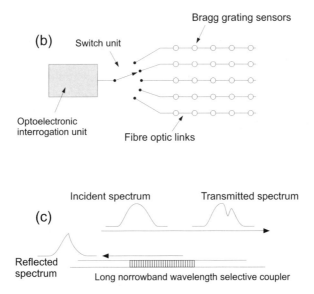

Fig. 2.8 The fibre optic Bragg grating – when operated as a sensor modulates the reflected wavelength: (a) basic grating geometry; (b) mode of operation – the grating reflects a precisely defined wavelength of light; (c) switched multiplexed network – each grating in a single string operates in a different wavelength range.

constant of the local structure, a direct measurement can be taken from variations in the grating reflected wave.

Multiplexing fibre Bragg gratings is in principle straight forward. A series of them operating at different wavelengths can be fabricated along the length of the fibre. It is then simply a matter of injecting the broadband light source into the fibre and measuring the reflected spectrum in which a particular wavelength band corresponds to a particular sensor. A fibre optic commutator switch can be used to address several of these chains of sensors and arrays of up to many hundreds have been demonstrated based upon this simple architecture.

The benefits of the fibre Bragg grating are substantial. The grating period is coded into a reflection wavelength which is an unambiguous signal and is independent of illuminating intensity or detector sensitivity. The fibre with the gratings within it is easily attached to or embedded within the structure and a single simple operation installs the entire sensor array. There are of course some disadvantages. The interrogation scheme requires precise wavelength calibration and measurement. Sometimes this is provided by a reference temperature stabilised, strain isolated grating, sometimes by an independent laser source for example a helium neon operating on a near infrared line. The wavelength stability of the reference should be better than ± 0.1nm over the environmental operating range of the receiver. The wavelength decoder itself can be based upon a number of concepts though a scanned Fabry–Perot interferometer is probably the most popular. Variations on this theme include tuned acousto-optic detection systems and dispersive interferometers illuminating CCD arrays. None is nearly so simple as a strain gauge amplifier and the wavelength decoder remains a major factor in the somewhat high costs of FBG systems. There is also an operational issue concerned with temperature compensation and the seriousness of this depends upon the accuracy of the strain gauge measurement which is required. In broad terms 1°C temperature change gives the same wavelength shift as 10μstrains change in strain. Additionally for most applications involving static strain measurement a temperature map is a necessary complement to the strain map since it is essential to determine whether a change in strain is due to a change in temperature or due to an applied load. Mounting FBG sensors, and by implication the design of their packaging, is inherent within the procedures required to correct for these discrepancies. The packaging process must protect the grating from damaging moisture ingress and also ensure intimate contact between the grating and the structure. Additionally it must be compatible with on-site handling in the application context. Progress in packaging has been relatively modest since in most demonstration applications the FBG has been attached to or embedded within the structure by hand rather than via a defined procedure.

That said a few real applications have emerged. FBG sensors have found their way into the composite mass of very expensive racing yachts where they are embedded within the root of the mast deep in the composite material to ascertain the strain history. They have been tried out in countless bridges and dams, to measure the drift in the foundations and superstructure of ageing buildings and in the hulls of high speed marine vehicles based upon composite manufacture.

There are also other novel fibre optic approaches used in sensing for smart systems and again the most important ones use the specific features of fibre technology which enable multiplexed, distributed or integrated measurements. Distributed measurements facilitate monitoring the measurand as a function of position along the length of an optical fibre with spatial resolutions typically of the order of one to a few metres. The range over which these measurements may be made can extend up to several tens of kilometres and this is a truly unique feature of optical fibre sensor technology; no other measurement concept offers this capability. Integrated measurements effectively present the average of a distributed measurement function over the length of the optical fibre — these incorporate the line integrating capabilities of the fibre and, whilst this is not unique to optical fibre measurements, the total access range over which such integration may be achieved and the range of measurands to which it can apply are also unique to fibre technology.

For smart systems the applications for such specialised measurements are many. However all rely upon the need for access over long range or the need for measurements involving relatively long gauge lengths. To illustrate this we shall look at three examples.

In civil engineering integrated strain over gauge lengths of metres give us extremely useful smoothed information concerning structural behaviour. A strain gauge, or fibre Bragg grating in contrast, measures a local event which may distort the global perspective since cracks and very local damage are quite common in civil structures. The SOFO system is a tried and tested fibre optic measuring probe which has proved capable of stabilities of the order of a few microns over gauge lengths extending to several tens of

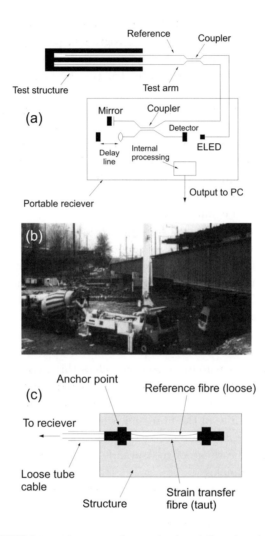

Fig. 2.9   The SOFO integrating sensor for monitoring civil engineering structures: (a) overall system diagram; (b) working on site in Switzerland; (c) the sensor itself.

metres. Several thousand of these sensor systems have now been installed predominantly in structures in Europe and numerous detailed case histories of these are available (Fig. 2.9). These cover applications ranging from monitoring the impact of the new load imposed by a bridge extension on a major highway, to examining the progress of cracks in ancient churches, to assessing the integrity of security of dam construction procedures. The system works on a very simple and well known optical principle, namely, white light interferometry. This enables path difference measurements between the coupled and temperature compensating arms of the fibre interferometer to be made with accuracies of less than one fringe. The measurement drift observed, which corresponds to several fringes, arises predominantly in the mechanical reference system and in the fibre mounts within the interferometer itself. Measurements made over time spans of several years have however confirmed the overall stability of the system as just a few microns. The SOFO system has established itself as an extremely reliable reference measurement technique.

Stimulated Brillouin scatter is a nonlinear effect and sounds very esoteric as a fibre measurement system. Stimulated Brillouin scatter however precisely couples a forward travelling optical wave into a reverse scattered Brillouin shifted wave with a precisely defined relationship between the two. The reverse scattered wave is effectively reflected by an acousto-optic "Bragg grating" of which the period is precisely one half of the optical wavelength (Fig. 2.10). In other words the acoustic velocity at the back scattered Doppler shift in frequency is exactly that required to make the Doppler shift frequency have a wavelength of one half the optical wavelength. For most optical fibre operating in the near infrared this corresponds to a frequency shift of 12 to 15GHz. This frequency shift in turn — assuming we know the optical wavelength — gives a very accurate indication of the acoustic

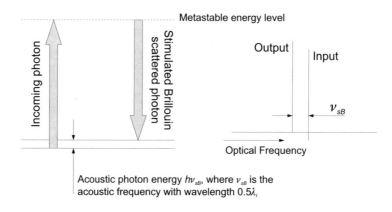

Fig. 2.10   Stimulated Brillouin scatter processes.

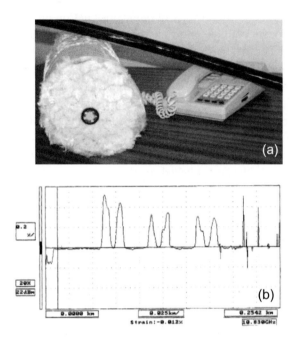

Fig. 2.11   OSCAR (optical sensing cable for ropes) – measuring the condition of marine ropes: (a) sensor cable in 1000 tonne rope model; (b) multiple sensor responses in experimental sensor cable.

velocity. The frequency shift can of course change as a function of position along the fibre so what is returned at the receiver is a map of acoustic velocity versus position — the optical wavelength remains constant. Elementary mechanics tells us that the acoustic (longitudinal wave) velocity is a function of Young's modulus, the density and the local strain, the first two of which are functions of temperature. We then have a system which returns effectively a convolved temperature plus strain map of the condition of the fibre. A particular strength of this system is that the probing distance can extend over many kilometres and this has many intriguing applications in for example monitoring cables especially in unstable ground, for example in earthquake zones. It has also found an application in measuring high tensile stressed marine ropes for use in high performance anchoring and towing systems (Fig. 2.11). In the former case the Brillouin probe enables the cable system user to take rapid corrective action before irreversible damage occurs. In the latter it enables the maximum possible use to be made of an

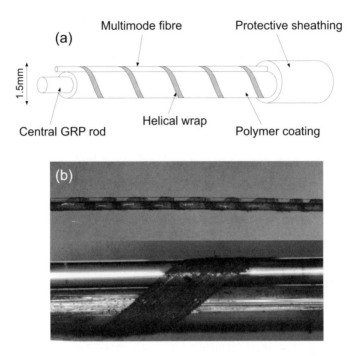

Fig. 2.12 Distributed sensor for water or hydrocarbon spillage: (a) schematic diagram; (b) microscope photographs.

anchoring system so that the very expensive tether replacement procedure can be implemented when required rather than on deliberately premature scheduled maintenance.

Distributed measurements based on backscatter find a diversity of other applications. The Raman probe temperature measurement is well known and provides an unambiguous temperature map along the fibre. This has found use in industrial processes and as a fire alarm system in tunnels.

Introducing a suitable transduction mechanism could also broaden the scope of distributed measurements into the chemical domain and a particularly interesting example of this facilitates monitoring oil spills and leakage from pipelines and storage tanks. In this case the transducer is a thin film of a specially tuned chemical which responds to the liquid to be monitored by physically swelling. This swelling process is then caused to introduce microbending loss along the fibre which can be monitored through time domain reflectometry (Fig. 2.12). Careful choice of the transduction material

— usually a polymer — its preparation and its adhesion to the substrate can give a chemical measurement system which is stable over several years and can be taken through many hundreds of wetting cycles.

Optical fibre sensor technology is then arguably the best suited to emerging applications in smart systems. This is primarily because of the ability for inherent multiplexing, either in distributed or Bragg grating architectures, coupled to the immense possibilities in the design of the interface between the structure to be monitored and the fibre which will do the monitoring. The technology has the additional benefit that it can be interfaced to numerous physical and chemical parameters through appropriately designed transducers though in all cases, in common with all other sensor technologies, some temperature compensation is almost inevitably required.

Fibre sensors will continue to address specific niche applications but again this is common to all sensing techniques. In particular applications which require large numbers of sensing points scattered over wide areas but compatible with a linear array will be especially attractive. The final arbiter is of course the cost:performance ratio coupled to the perceived value of the application concerned. The technical performance of the fibre sensors is, in most contexts which we have described here, extremely high but the cost typically matches the performance and assessing the value in a specific application is complex. However in most of the applications alluded to in the preceding discussion the value is sufficiently great to justify the relatively high cost of the fibre optic sensing system.

2.3.4 *MEMS*

The term "MEMS" refers to micro-electro mechanical systems. In the sensor context this includes the whole topic of micromechanical sensors.

Micromachining technology essentially modifies the photolithographic processes used in silicon manufacture into the means for realising mechanical structures. These photolithographic processes include not only the obvious mask based technologies but also techniques such as laser writing. Masking and photo-resist based processes define patterns on the surface of the machined structure. The machine is made three-dimensional by etching, usually exploiting a differential process. This differential process may, for example, attack different doping levels at different rates or may use the crystalline properties of the substrate material. Laser or indeed electron/ion

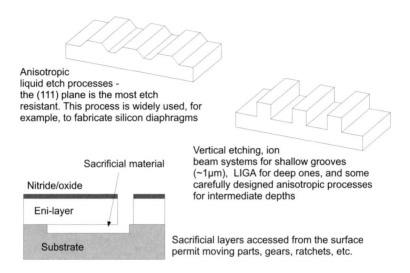

Anisotropic
liquid etch processes -
the (111) plane is the most etch
resistant. This process is widely used, for
example, to fabricate silicon diaphragms

Sacrificial material

Nitride/oxide

Eni-layer

Substrate

Vertical etching, ion
beam systems for shallow grooves
(~1µm), LIGA for deep ones, and some
carefully designed anisotropic processes
for intermediate depths

Sacrificial layers accessed from the surface
permit moving parts, gears, ratchets, etc.

Fig. 2.13   MEMS – basic principles of the major processes.

beam writing can be used to induce chemical reactions, to directly ablate materials or to provide another means for writing in resists. The tools are therefore many and can be further expanded through the use of wet and dry etches, photo assisted processes, differential stirring and mixing and a host of other ingenious material manipulation methods. The result is that three-dimensional structures can be fabricated though in the main these structures tend to be relatively large in two-dimensions (the surface of the substrate) and relatively shallow in a third (Fig. 2.13). There are though exceptions to this rule for example the LIGA process which uses nuclear radiation to define prismatic structures perpendicular to the substrate face in polymer materials.

MEMS emerged from silicon technology and not surprisingly most of the initial work in MEMS was done in silicon. Indeed much of the currently available micromachined artefacts are based upon silicon technology. The techniques are inherently very powerful. Silicon is mechanically excellent, stronger than steel with a high melting point and, in the single crystal form, readily available and remarkably stable mechanically. More recently polymer based MEMS structures and metal MEMS have become more important but to date practical applications have been dominated by silicon and so for the remainder of this short section we shall look at the type of sensing system which silicon has enabled. Incidentally in addition to the ability to manufacture very small mechanical structures it is clearly possible in principle to incorporate a local silicon micro circuit on to the same chip as the mechanical system. Whilst this has been done there are some processing incompatibilities between the chemical procedures involved in micromachining and those involved in circuit manufacture. Often the compromise is to place the circuit and the sensor within the same package but through separate bonding processes.

The sensors are mechanical but they should be interrogated electrically. Any sensing mechanism is therefore mechanical in origin. There are relatively few readily detected mechanical processes which can be realised in micromechanical form but there is a plethora of interfaces which can modulate these processes in order for them to be used as sensors.

The mechanical modulation mechanisms may be reduced to:

- The relative position of the components of the mechanical structure (i.e. displacements).
- The vibrational resonant frequencies of the structure.

These can be changed in response to either temperature or externally applied forces (Fig. 2.14). To my knowledge all the micromechanical sensors which have been discussed in the literature operate at the fundamental level in one or other of these two modes.

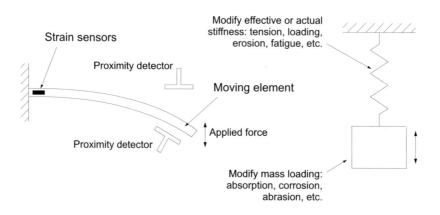

Fig. 2.14   Modulation techniques for MEMS sensors – position modulation and resonant frequency modulation.

Detecting the changes in relative position or in resonant frequency as an electrical signal is the second phase and again the detection process is dominated by two mechanisms. The first is to diffuse piezoresistors into high strain areas of the sensor and detect the changes in resistance in some sort of bridge circuit. This is probably the most straightforward technique and has been used extensively for well over 20 years. The second relies effectively on capacitive measurements between two planes, one of which is the moving plane. In some situations the motion can be detected optically which offers the prospect for a combination of optical fibre based interrogation and micromechanical fabrication. Whilst this may appear attractive in practice its use has been somewhat limited.

The first extensively used silicon MEMS based sensor was a simple pressure diaphragm using piezoresistors diffused into the high strain region at the edge of the diaphragm. A variation on this basic theme etches a mechanical suspended bridge resonator on to the diaphragm and uses the bending of the diaphragm to tension the bridge thereby changing its resonant frequency (Fig. 2.15). This affords higher accuracy and stability than the piezoresistive version and is still compatible with essentially the same process steps.

The basic resonant bridge approach can be used to monitor any parameter which will change the stress on the bridge. This of course can include temperature due to differential thermal expansion phenomena be-

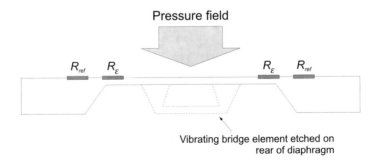

Fig. 2.15 Examples of MEMS using piezo resistance and vibrating element stress detection in a pressure diaphragm. The resistors R_{ref} and R_ϵ are reference and strain sensitive respectively and form a bridge network. The bridge is strained as the diaphragm responds to the pressure. Its resonant frequency is also temperature sensitive. The resistors are temperature compensated.

tween the various components of the sensor and its housing and also due to
the changes in material parameters — notably stiffness — with temperature. Careful design can minimise or even eliminate temperature sensitivity
over wide thermal excursions so that the focus can move to examining other
physical and chemical phenomena. The stresses on a microstructure of this
nature can be varied by changing the gas flow incident upon the bridge, by
coating the bridge with an active chemical which attempts to change its
physical dimensions in response to chemical stimuli (changing the tension
of the bridge) or by absorbing selectively chemical materials and thereby
changing the mass of the vibrating element.

A proof mask can be machined into the centre of the bridge structure
and this proof mask can itself then form the basis of a vibrating element
accelerometer — the force which the proof mask exerts on the centre of the
bridge depends on the acceleration perpendicular to the direction of the
bridge (Fig. 2.16). The accelerometer can be made directional by arranging
for the bridge to be much stiffer in the direction orthogonal to the sensitive
axis. Accelerometers can also clearly be realised in a force feedback mode
using a simple cantilever or even in a displacement measurement mode for
an open loop system since silicon is so very linear and exhibits very little
mechanical hysteresis than even an open loop system can perform very
adequately.

Micromechanical gyroscopes have also attracted attention. These in effect detect centrifugal forces (often through the Coriolios force). This force
can, for example, be used to modulate the resonant frequency of a rotating
tuning fork structure (the tuning fork is rotated at the rate which is to be
measured) in ring structures or in wine glass structures. In ring and wine
glass structures the Coriolios force causes a differential change in the resonant frequencies in two directions around the loop thereby changing the
rotation rate signal into a differential resonant frequency. One of the many
attractive features of gyroscopes based on this technology is that there are
no rotating bearings so that an all solid state configuration is eminently
feasible. The wine glass gyro has been researched for at least 25 years but
as yet has made modest impact primarily because of the very tight machining tolerances necessary in its manufacture to achieve good sensitivity.
The micromachine gyroscope in contrast can be produced with demanding mechanical tolerances at the rate of hundreds or thousands per silicon
wafer. The sensitive mass is however very small so that the rotational rate
of sensitivity is somewhat limited. Accordingly, despite their availability,

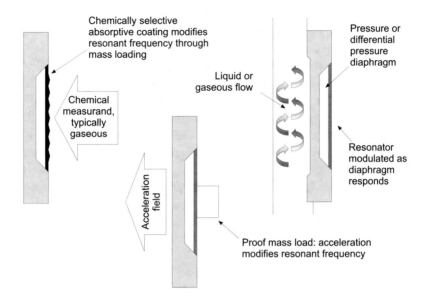

Fig. 2.16 Modulating the resonant frequency of a bridge – forces due to flow, acceleration, chemistry, etc.

the ingenious silicon micromachined gyroscope has yet to make a major impact.

In the smart structures context MEMS offers largely orthogonal application prospects to fibre optics. MEMS sensors are small and can be precisely located. A number of sensors targeted towards different measurands can in principle be realised on a single chip. This particular feature of MEMS technology has, to my knowledge, been but rarely investigated despite the obvious potential in, for example, in vivo medical applications. MEMS sensors are potentially very inexpensive and their cost is usually dominated by packaging, test and certification procedures. They are very compatible with high volume applications exemplified by their use as the accelerometer in probably every single automobile airbag system world-wide.

For smart systems their penetration has been very modest, perhaps because their niche has yet to be accurately identified. MEMS will work at high temperatures — provided the silicon need not be a semiconductor at the same time. MEMS can offer dimensions of a small fraction of a millimetre for an entire sensing element. MEMS systems can be interrogated optically or electrically through umbilical links a small fraction of a millimetre in diameter. The mechanical structures are small, can respond with quite surprising speed with time constants into the microsecond area and below, and are physically non-intrusive.

It would seem then that MEMS sensing, in contrast to fibre optics, is compatible with small structures rather than large ones. Smart systems for near field probes are already in use. There may be scope for MEMS in machine tools for very high precision. There is certainly scope for MEMS in biomedical smart systems. The relationships between the application potential for MEMS and the apparent or actual applications have yet to be made. MEMS sensors are a tool awaiting their niche but the niche will emerge as the needs for precision machining and manipulation in bioscience, medicine, environmental control and precision mechanical manufacture become more pressing.

2.3.5 *Piezoceramics and Piezoelectric Polymers*

Sensing using piezoceramics and piezoelectric polymers involves almost exclusively using the very simple properties of piezoelectric material, a pressure or strain on the material is converted to a change in electric charge distribution across two faces of the material which in turn — effectively through the capacity effect — appears as a change in voltage (Fig. 2.17). Piezoelectric materials are dominated by piezoceramics using lead zirconate

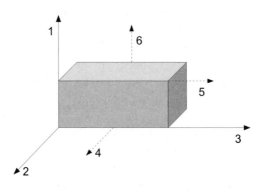

Fig. 2.17 Basics of piezoelectricity: stress axes (1, 2, 3) and shear planes (4, 5, 6). Stresses, shear or axial, in general produce electric fields in all axial directions, linked via the piezoelectric strain coefficient tensor. Usually, the E field along the stress dominates, but not necessarily.

titenate as the dominant host material. Piezoelectric polymers — almost invariably PVDF — are the second most utilised of the piezoelectric sensing systems and finally some piezoelectric crystals including barium titenate, lithium niobate and zinc oxide fall into a third class of material structures.

Interestingly piezoelectric sensing systems (which can also be used in the reverse direction changing an electric field variation into a mechanical stress variation and then usually into an ultrasonic wave) have also found their way into both MEMS and fibre optic sensor technologies. Piezoelectric polymers have been coated on fibres to form electric field sensitive devices and the number of materials — predominately zinc oxide — have been used as both mechanical (ultrasonic) sources and sensors integrated into MEMS processing. Crystalline piezoelectric materials such as lithium niobate and some piezoceramics have also been used to excite and detect ultrasound in relatively large scale MEMS structures — almost invariably for sensing.

In the smart systems context piezoelectrics may potentially fulfil one of three roles (Fig. 2.18):

- As point sensors of pressure or strain, especially for ceramic and crystalline piezoelectric materials.
- As integrating area sensors, particularly piezopolymers.
- Tactile or pressure distribution sensing by using an interconnect pattern to address tactile pixels across the area of a polymer film.

Whilst piezoelectrics are undoubtably of interest and concepts such as the tactile pressure pattern system, whilst attractive, have yet to make a significant impact.

2.3.6 *Film Technologies: Coatings and Threads*

Structural sensing using coatings has been with us for quite some time and coatings which change colour in response to external stimuli are well known. Usually these coatings and their equivalents in woven thread formats find it difficult to give the user accurate quantitative information. However qualitative information is often all that is required and in some cases a suitably interpreted and average array of qualitative indicators can actually give a quantitative result. Films and threads are, in the engineering context, "bad sensors" but in essence are how most biological systems obtain their information — with the possible exception of the visual and auditory functions.

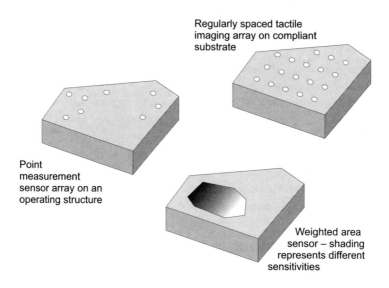

Fig. 2.18 Piezos as sensors for point, integrating area and tactile distribution.

Perhaps the best known of the external film based technologies is the sweatshirt which turns red as its temperature approaches body temperature. These have been the source of many a — not necessarily too polite — comment. In a more technical context paints and coatings have been used to indicate impacts on surfaces ("bruising paints") and to provide a permanent record of a change in temperature above a particular threshold value. Most — possibly all — of the so-called smart paints involve polymer chemistry to produce pigments which change their visual properties irreversibly after thermal or physical stress.

Fibres in contrast almost invariably invoke some form of strain:resistence dependence. The carbon fibres in CRP are piezoresistive. This has been used to good effect to monitor the strain levels — albeit rather qualitatively — in CRP structures. Using the reinforcing fibres as a monitoring system has obvious attractions despite the fact that this is not a "good" engineering sensor since its response cannot be guaranteed from one structure to another. However it is certainly a good qualitative sensor system since trend analysis on the results can give a reliable indicator of the underlying structural condition.

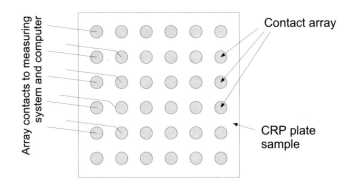

Fig. 2.19 Using carbon fibres in CRP as a resistance based damage sensor.

Resistive polymer threads allow the same basic idea to be extended into flexible as well as rigid structures. It is arguably here that the potential in the smart structures context becomes much more intriguing. The polymer threads can be woven into sheets and these sheets can be readily attached to structures of any shape and with an array of electrical connections to the ends of the fibres we can realise a stick-on condition monitoring system. Further the stick-on system can also be embedded into some types of composite material — particularly GRP — where the non-conducting fibres preclude the use of the direction approach exemplified in CRP but the fabrication process is cold so that the polymer sensing mat would not deteriorate during fabrication.

These sensors are however "biological" in their basic function. Unlike the nervous system based on fibre optics or MEMS technology, where each sensor element is repeatable and calibrated, our woven array is entirely different and must be taught to respond appropriately. Of course given modern artificial intelligence programming systems learning for a computer — at least in this context — is a relatively straightforward if still somewhat tedious operation. However one could conceive of in-service learning rather than a pre-service calibration and of auxiliary sensing systems to provide spot calibration to check and update the learning as appropriate. It is in this area that perhaps the concept of smart structures offers up the most potential benefit. It is also an area which has been but very little explored.

Applications for such sensitive mats and coatings are enormous. The virtual reality glove is an obvious one. Medical analysis by slipping on

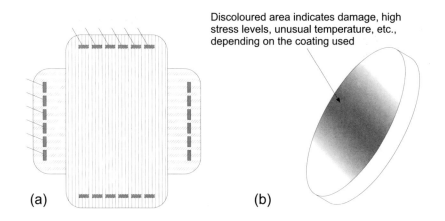

Discoloured area indicates damage, high stress levels, unusual temperature, etc., depending on the coating used

(a)

(b)

Fig. 2.20 (a) 'Smart' weaves for semi-quantitative surface and/or internal condition monitoring; (b) 'smart' paints for semi-quantitative surface condition monitoring.

a leotard or a strap rather than attaching electrodes could be another. Actuator systems could be built into intelligent prosthesis and could assist in the curing of injured joints and limbs. This entire area using poor sensing in an intelligent context is one which offers great potential for the future (Figs. 19 and 20).

2.4 Conclusions

Sensing is the starting point in the smart structure. Historically we have attempted to engineer the sensor to give precise data. The first four subsections in part three have focused on examples of this quest for precision.

However the availability of ever more computing power for ever less investment does offer the prospect of trading off sensor precision and reliability for learning and averaged intelligent estimation — following the concepts so effectively exploited in biological systems.

At this stage it is difficult — probably impossible — to quantify how effective this trade-off may prove to be in the evolution of future smart systems. However it certainly is the case that no matter how the smart system is designed and realised it is far preferable to view the sensors in a system context than simply as isolated components. Indeed an integrated approach to engineering the entire artefact — be it smart leotard or suspen-

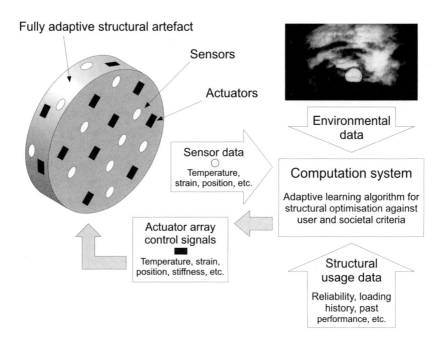

Fig. 2.21 The smart structure incorporating learning and response functions.

sion bridge — is essential and as engineering design progresses the concepts of structures, sensing, actuation and intelligence must merge into a simple set of operational criteria.

This chapter has attempted to present an overview of current approaches to sensing to give some impression of the importance of signal processing in sensor systems and to speculate how sensor technology may evolve as the processing and the sensors become increasingly integrated. We can anticipate intriguing developments as this process evolves.

Chapter 3

Vibration Control Using Smart Structures

Michael I. Friswell[1] and Daniel J. Inman[2]
[1]Department of Mechanical Engineering,
University of Wales Swansea,
Swansea SA2 8PP, UK.
[2]Center for Intelligent Material Systems and Structures,
Virginia Polytechnic Institute and State University,
Blacksburg, Virginia 24061, USA.

3.1 Introduction

All structures vibrate to some extent. Usually in aerospace, robotics and automotive structures this vibration is not wanted and can lead to catastrophic failure. This failure may arise in the long term due to fatigue causing cracks, or it may occur quickly due to the material not being able to cope with the material stresses from the large amplitude displacements. Even if failure does not occur the resulting noise and vibration is often unpleasant, and certainly not a desirable feature in a consumer product. In many applications the structure may be stiffened so that the response is less, and the resonances of the structure are moved above the forcing frequencies. Alternatively discrete dampers similar to automotive engine mounts or shock absorbers may be used. However these are large and heavy devices, and only a limited number of them may be used. The trend in aerospace and automotive applications is to produce lighter structures that are more energy efficient and responsive. This not only makes the structure less stiff, and therefore more susceptible to vibration, but also makes controlling the

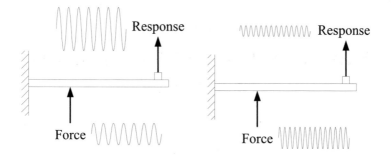

Fig. 3.1   Response of a structure to a sinusoidal force.

vibration more difficult. This chapter highlights some of the techniques that may be used to reduce the vibration of a lightweight structure, and introduces the concept of a *smart structure*.

But what is a smart structure, and how can it be used to control unwanted vibrations? Essentially a smart structure is one where the means of controlling the vibrations is incorporated within the structure itself in an unobtrusive way, rather than adding bulky external sensors and actuators. The sensors measure the response of the structure, and after suitable processing produce electrical signals to drive the actuators to generate forces to counteract the vibration. A substantial amount of research effort has been undertaken in the area of smart structures and intelligent material systems over the last decade or so. This chapter can only highlight the main results of this research, and give some typical examples. The references should be consulted for more detailed information.

Before introducing smart structures some concepts from vibration and dynamics, such as resonances and mode shapes, are introduced. Such background is vital to understand the nature of vibrations, and thus the means by which they may be reduced. This section is only a basic introduction and Inman [15] or Avitabile [1] should be consulted for further information. Section 3.2 considers typical sensors and actuators that may be used. The most popular for both transducers are piezoelectric components, and these are discussed in detail. Other types of sensors and actuators may be used, and the interested reader is referred to the other chapters in this book for more information on these technologies. There then follows a review of some of the common methods of control and their application. Finally some examples of simple smart structures are given in Sec. 3.4.

3.1.1 *The Dynamics of Structures*

Anyone who has felt the oscillatory response of an automobile or washing machine should understand that structures vibrate. The process that produces this response is that an oscillating force (due to the automobile engine, or the rotation of out-of-balance washing), causes the structure to respond. Amazingly, if the forcing is sinusoidal (that is at a single frequency), then for a linear system the response will also be sinusoidal at the same frequency as the excitation. We will only consider linear systems, which covers a wide range of aerospace, automotive, and other structures, particularly when the response magnitude is small. Figure 3.1 shows the force and response diagramatically. So what happens if the frequency of the excitation changes, while the peak force level remains unchanged? Well, the response is now at the new excitation frequency, and the magnitude of the response will change. Figure 3.1 shows the effect of two different frequencies applied to the same structure.

Suppose now, we note the magnitude of the response at a large number of different excitation frequencies, and plot the results on a graph. A typical example is shown in Fig. 3.2, and is referred to as the frequency response function (FRF) of the structure. The peaks in the FRF are called resonances, and the frequency where these resonances occur are close to the *so called* natural frequencies of the structure (in fact resonances occur at the natural frequencies if damping is neglected). If the structure is excited with a force whose frequency corresponds to a natural frequency, then a

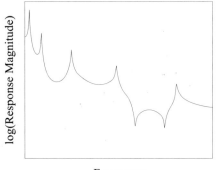

Fig. 3.2   Frequency response function of a structure.

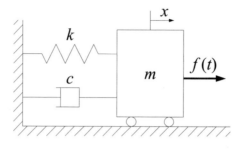

Fig. 3.3   Single degree of freedom discrete system.

small force can produce a large response. These are frequencies at which the structure naturally wants to vibrate, and indeed if a structure is deflected statically, and then released, the response will consist of a sum of responses at the natural frequencies.

Damping causes the energy within a vibrating structure to dissipate, and eventually the motion will stop. Damping arises from many sources, and is very difficult to model. For analysis purposes viscous damping, where the magnitude of the damping force is proportional to velocity, is very often used. An automotive shock absorber acts approximately as a viscous damper. Figure 3.3 shows a simple single degree of freedom system consisting of a mass, spring and viscous damper. The number of degrees of freedom refers to the number of independent co-ordinates required to determine the unique position of the structure (in this case one). Figure 3.4 shows the FRF and time response of this system for three different values of damping. Increasing damping reduces the magnitude of the response at resonance, and also causes the time response to decay faster. In vibration control, increasing the damping is vital. The level of damping is usually given using the damping ratio, which is the ratio of damping coefficient to that required for the system to be critically damped. Critical damping is where the response just ceases to be oscillatory, as shown by the highest level of damping in Fig. 3.4.

3.1.2   *Modal Analysis of Structures*

The next important question is how the structure is vibrating at the natural frequencies of the structure. Figure 3.5 shows the FRFs measured

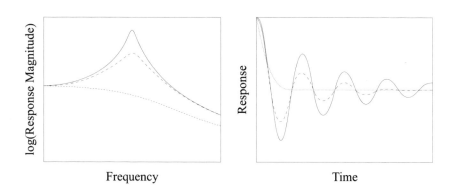

Fig. 3.4   FRF and time response of a single degree of freedom system for damping ratios of 10% (solid), 25% (dashed) and 100% (dotted).

at different locations on a cantilevered beam (i.e. a beam that is fixed at one end) close to the first three resonances, due to an excitation at a fixed location. Each thin line shows the magnitude of an FRF for a particular measurement location. The FRF magnitude is shown as negative depending on whether the response at a frequency just below resonance is in the same direction as the force or in the opposite direction. The thick line on the plots joins the points of maximum response for the individual FRFs. These deformations at resonances are called mode shapes (again this is only approximately true, but it serves our discussion, and in practice the approximation is very close). The modes shapes encode the spatial motion of the structure. The modes of a structure, determined by its natural frequency, damping ratio and mode shape, are a vital building block. The simplest system consists of a mass together with a spring and dashpot connecting the mass to ground, as shown in Fig. 3.3. This idealized system has just one natural frequency and one resonance. It may be demonstrated that the response of a general multi degree of freedom structure is actually a sum of responses in each mode, and the response in each mode is equivalent to the simple single degree of freedom mass, spring and dashpot system.

There are three important steps in performing a modal analysis of a structure. The first step is to transform the force that arises physically on the structure, into a force into each mode. Here the mode shapes are very useful. If a structure is excited at a node of a particular mode (the displacement at a node is zero), then the mode in question will not be

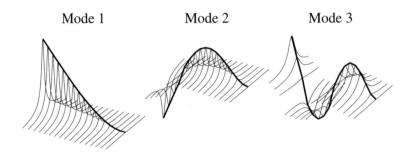

Fig. 3.5   Modes of a cantilever beam.

excited, i.e. there is no force into the single degree of freedom modal system. Similarly, if the force is applied at an anti-node of a mode, where the mode displacement is maximum, then the mode is excited with the greatest force possible. Figure 3.6 shows this for a cantilever beam. Here the response is measured at the tip of the cantilever beam, and the beam is forced either close to the clamp (solid) or close to the node of the second mode (dashed). Clearly when the beam is forced near the node of the second mode, this resonance is not excited and is not present in the response.

After the force is transformed into the modes, the response of the equivalent system for each mode is derived. The response in each mode is then summed, using the mode shapes, to obtain the actual response of the structure. Clearly the response at a node of a mode shape, will not show any evidence of the presence of the mode in question, as shown in Fig. 3.5. The whole process of this *modal decomposition* is shown graphically in Fig. 3.7.

3.2   Sensors and Actuators

Active structures measure the response of a structure, process these measurements and apply forces to counteract the unwanted vibration response. It is clear that the transducers used to measure the response (sensors) and the means to apply the force (actuators) are vital components. Indeed the requirement of a smart structure that the transducers are capable of being embedded into a structure make further demands on the choice. Although many different sensing and actuating mechanisms exist that may be utilized in a smart structure, the emphasis here is on piezoelectric transducers. These components may be easily incorporated into structures, and are

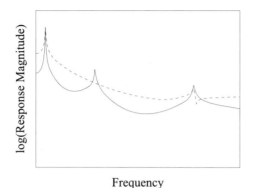

Frequency

Fig. 3.6 Effect of force location on the response of a cantilever beam. The force is located at a quarter length (solid) and three quarter length (dashed).

relatively straightforward to use. The other chapters in this book should be consulted for more detail of different sensor and actuator concepts.

A piezoelectric material produces a electric charge when it is mechanically stressed by an external force. Furthermore, if a voltage is applied to the material its dimensions will change. Thus the same material may be used as a sensor or as an actuator. The way in which the material responds electrically to mechanical forcing, and vice-versa, depends on the dimensions of the element, and the actual material characteristics (in particular the coefficients representing the mechanical–electrical coupling). The main problem in using piezoelectric material as an actuator is that the mechanical strains are small. The best approach is to use thin sheets of piezoelectric material, off the neutral axis of a plate or beam structure, so that the strain in the actuator causes the plate or beam to bend. This is shown diagramatically in Fig. 3.8. A conductive coating is applied to the upper and lower surfaces of the piezoceramic sheet in order to generate the electric field through the sheet. Actuators are usually made from ceramic piezoelectric material, and sensors can be made from either ceramic or polymer film.

Thus far, this section has considered the use of piezoelectric components for sensors and actuators. For control purposes placing the sensors and actuators at the same place on a structure (a collocated sensor–actuator pair) has many desirable consequences, most notably that ensuring the controlled system is stable for all modes (that is over a wide frequency range) is simplified. For beam and plate structures it is possible to mount the sensor

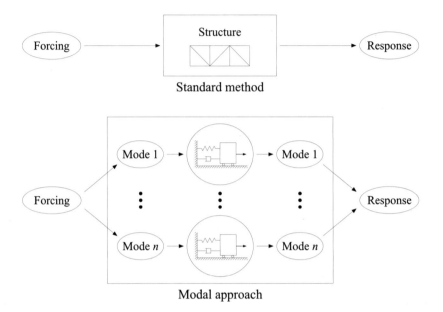

Fig. 3.7   Modal approach to the analysis of structures.

on one side of the panel and the actuator on the other. Alternatively the sensors and actuators may be embedded into different layers of a composite structure. A different approach consists of using a single transducer element as both a sensor and an actuator, the *self sensing actuator* (see Dosch *et al.* [3] for details). The self sensing actuator circuit allows each piezoceramic patch to behave simultaneously as a sensor and an actuator

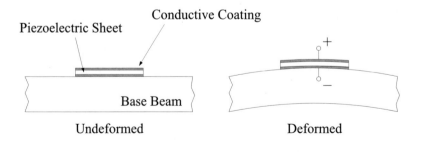

Fig. 3.8   Action of a PZT actuator.

by essentially using a bridge type circuit to keep track of that part of the voltage across the piezoceramic due to the control input voltage and that part of the voltage due to sensing.

3.3 Active Control of Structures

The previous sections have introduced the concept that a structure has complex dynamic properties, and also that sensors and actuators are required to enable a controller to interact with the structure. The sensors measure the response of the structure, and the actuators enable the application of arbitrary forces. If we apply these forces in response to the sensor measurements we have a *feedback* control system, where the sensor measurements are fed back to generate the control forces. The forces then cause the structure to respond, producing an output at the sensors. Because of this the system is called a *closed loop* control system. Without this feedback control the system is *open loop* (since the loop is not closed).

A simple example of a dynamic system is a car driving along a straight road. The sensors are the eyes of the driver, that judge the position of the car on the road. The actuator is the force applied to the steering wheel. In open loop the steering wheel will be set so the wheels are straight, and providing the car's initial direction is correct the car will continue along the straight road. However, it is unlikely that the car's initial direction will be exactly correct. Furthermore, the road maybe slightly inclined, a cross wind may be present, and in practice a road is never straight. These problems mean that without control the car will quickly leave the road. The controller in this case is the driver. He senses the position of the car, and turns the steering wheel to ensure that the car remains on the road. This closed loop feedback control system enables cars to be driven over roads with bends, and subject to environmental forces that apply disturbance forces to the car.

But how much should the steering wheel be turned for a given error obtained from the sensors? This ratio of control force to sensor output is termed the *gain* of the control law. If the steering wheel is only turned a small amount (the gain is low) it will take a long time for errors in the car position to be corrected. For larger gains the system response often overshoots the desired position, which is then corrected, leading to an oscillatory response. Figure 3.9 shows a typical response to a step change in

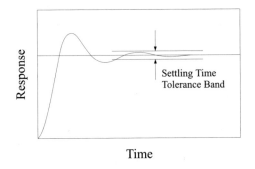

Fig. 3.9   Step response of a typical closed loop system.

the demanded position of the system (e.g. the direction in which the car
is heading is changed). The response is equivalent to the single degree of
freedom discussed earlier. Some overshoot is desirable, since this ensures
that the speed of response of the system is relatively fast. A measure of
control performance is the *settling time* which measures the time taken for
the transient response due to a step change in demand to decay to a given
percentage (usually between 2 and 5%) of the steady value. If the gain
is increased further there usually reaches a point where the control over-
compensates for the measured error, and actually makes the oscillations
grow. This is shown diagrammatically in Fig. 3.10, and the system is *unstable*. An unstable closed loop system is clearly undesirable and should be
avoided at all costs.

3.3.1   *Modal Control*

There are a number of important criteria in the performance of an active
control system. It is vital that the sensors measure the response of the
structure accurately and that the actuators are able to apply a suitable
force to the structure to control the response. These simple ideas lead to
the concepts of observability and controllability [14]. Often the control and
analysis is considered on a mode by mode basis, and usually it is required to
control the response of a range of modes. In Sec. 3.1.2 it was demonstrated
that if the response was measured at a node of a vibration mode shape then
the mode in question will not appear in the response. Therefore the mode is
not observable. Increasing the number of sensors used to measure the mode

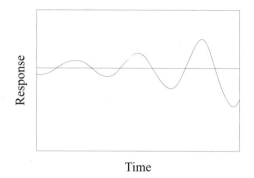

Fig. 3.10   Response of an unstable system.

means there is more chance of one of the sensors picking up the response of the mode. However, if a beam were bending in the horizontal plane, then this will not be observed by sensors measuring the response in the vertical plane. Similarly, if a structure is forced at a node, then no energy will be transferred into the corresponding mode. It is therefore impossible to alter the damping ratio of this mode, or to dissipate energy from this mode. Therefore the mode is not controllable using this actuator. Once again using two or more well placed actuators helps to avoid uncontrollable modes. It is clear from the above discussion that the mode shapes have a vital role to play in determining the controllability and observability of the closed loop control system. The above discussion has dealt with the cases where a structure is not controllable or observable. If a sensor was placed near a node of the mode shape, then the contribution of the corresponding mode to the measured response will be small, and therefore the mode will be barely observable. Thus, in practice, measures of observability are used that show how close a mode is to being unobservable. Similar measures may be developed for controllability.

The power of the concept of modes means that the control performance is usually considered based on its effect on the structure's modes. Often the response of the sensors is transformed to a set of modal responses, and the control effected on the modal response. The modal control forces are then combined to produce the actual physical force applied to the structure. The modal response may be obtained by combining the response of a number of sensors, weighted according to the mode shapes, or by filtering in the frequency domain. An alternative to having many discrete sensors, is to have a single shaped sensor, usually made from piezoelectric film, that provides the spatial filtering within the sensor [7].

Once the sensors and actuators have been placed, it remains to determine the algorithm to generate the control force from the structure's response. If the transducers are grouped as sensor–actuator pairs, then one option is to provide a controller locally, using one sensor and one actuator, or a single-input single-output (SISO) controller. The alternative is to combine the response of many sensors, to give the required force at many actuators, or a multi-input multi-output (MIMO) controller. Clearly a SISO controller is much simpler than a MIMO controller, although the latter, if well designed, should perform better.

3.3.2 *Adding Damping — Derivative Feedback*

In principle the simplest SISO controller replicates the viscous damping mechanism. If the sensor and actuator are collocated (that is located at the same position) then if the derivative of the position (or strain) response is negated, amplified and applied to the actuator as a control signal, the effect is exactly the same as a physical damper. Negating the signal is important, since the actuator resists the motion of the structure. Damping is added to all the modes. Increasing the gain of the amplifier increases the damping. Instabilities can arise with non-collocated sensor–actuator pairs, if for some modes the applied force does not resist the response. This can happen if the sensor and actuator are located at positions where the sign of the mode shape are opposite.

Adding Derivative (D) feedback increases the damping. If the displacement is also fed back to produce a force proportional to displacement, then the stiffness of the closed loop system is increased. This is a Proportional–Derivative (PD) controller.

3.3.3 *Positive Position Feedback*

The uncertainty in the parameters of a layered material demand that the closed loop control system be chosen to be robust to parameter changes, yet allow a substantial amount of damping to be added to the structure. Such a control scheme is provided by a form of independent modal space control [19] known as positive position feedback (PPF) [11; 6].

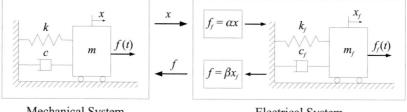

Fig. 3.11   Schematic of the PPF controller.

The PPF approach starts by assuming that the structure can be represented as a series of independent, decoupled modal equations, as discussed in Sec. 3.1.2. Thus the response of the structure is split into the contribution from each mode. The response within each mode is equivalent to a spring and dashpot system of unit mass, and the controller produces the force that controls the vibration of this single degree of freedom system. The force required for each mode is then applied to the structure using the physical actuators. We will consider the PPF control of a single mode. PPF control introduces a second single degree of freedom system in the electronic circuitry. This compensator is equivalent to a physical mass, spring and dashpot system, and is *forced* by the modal response of the actual structure. The modal force to drive the physical actuator is proportional to the response of the compensator system. The constant of proportionality used is the filter stiffness, which is positive, and hence the name Positive Position Feedback results. This is shown diagramatically in Fig. 3.11. In this way the physical mode and the compensator are coupled, and the damping in the compensator may be increased to damp out the physical vibrations.

Note that the filter natural frequency and damping ratio are chosen electronically by the closed loop design process. Also note that the control filter requires the measurement of the modal position coordinate. As long as the filter frequency is chosen to be below the structural frequency it is tuned to, the result is a stable controller, regardless of the values of the damping ratios. These conditions result in a number of practical observations.

For stability only knowledge of the structural frequencies, is required, and no knowledge of mode shapes or damping ratios is required. From testing principles (see Ewins [5] or Inman [15] for instance), the only structural property that can be consistently measured with any reasonable precision are the structure's undamped frequencies. Damping ratios and mode shapes are very difficult to measure. They are also the only parameters affecting the closed loop response that can be extracted with any accuracy from analytical models. Hence, the conditions for stability are very practical.

Because the physical and the electronic single degree of freedom systems are statically coupled, increasing the filter damping ratio also increases the energy dissipation in the structural modes. Furthermore the filter damping ratio can be increased without regard for loss of stability. Hence this form of compensation adds damping to the structure.

A potential flaw in the use of PPF filters, is that of assuming that the structural modes decouple. This concern is somewhat reduced by the fact that the PPF filter response rolls off at high frequency, so that the high frequency modes are not excited. None the less, some care should be exercised. A comparison between PPF and other popular control laws is given by Leo and Inman [16] and Dosch *et al.* [4].

3.3.4　*Other Controllers*

The number of possible controllers that may be applied to the vibration control of smart structures is vast. Here we concentrate on a typical approach, namely the Linear Quadratic Regulator (LQR). This method assumes that the control force is a linear combination of the states of a system, and uses optimization to find the best values of these gains. The states of the system are those quantities required to uniquely describe its configuration, and usually comprise the displacements and velocities of the structure. The number of states required is often quite high, and the great advantage of smart structures is that a large number of sensors may be used, thus making the job of estimating the state much easier. The performance index that is optimized in an LQR controller is the weighted sum of the output response and the control force required. Thus the result is a compromise between the response (which we would like to be zero) and the energy required for the control.

3.4　Examples of Vibration Control

This section gives typical examples of vibration control of smart structures. The initial example is possibly the simplest possible, namely a cantilevered beam in bending, with either a surface mounted or embedded piezoceramic actuator. The following examples are successively more complex configurations. In particular, four applications are discussed which point out natural and perhaps unique solutions to vibration suppression problems provided by a smart structure approach. The first of these examples consists of the slewing motion of a flexible beam through its bending direction around a rigid hub driven by a motor. Such motions are common in space and robotics applications. The addition of a piezoceramic based closed loop system is shown to significantly impact the power and performance of the slewing configuration.

The second application presented examines the vibration suppression of slewing frames, dynamically similar to those used as solar panels on satellites. Such frames are rich in coupled bending and torsion motion and known to vibrate excessively while slewing. The torsional motion is not able to be suppressed by the use of the motor alone. Here piezoceramics mounted directly on the frame are shown to render the torsional motion controllable providing an order of magnitude improvement in system performance. Thus a smart structure approach is shown to provide a solution to a difficult vibration suppression problem not solvable by conventional sensors and actuators. Both theoretical prediction and experimental verification are presented. Power consumption is shown to be minimal.

The third application examines the vibration suppression of a ribbed antenna similar to those used on satellites. Such structures exhibit repeated and nearly repeated natural frequencies. Hence, controllability becomes an issue and again a smart structures approach provides a low cost natural solution to a practical vibration problem.

The last application is that of a flat plate, commonly used to study aircraft or automotive components, clamped into a standard test fixture. Such plate elements are currently damped using constrained or free layer damping treatments in both aircraft (fuselage panels) and automobile applications (floor panels). The difficulty with using passive damping treatments based on viscoelastic materials is weight, environmental considerations and temperature dependence. The application of smart materials for controlling plate vibration has the potential for solving all three of these problems.

These results indicate a clear, logical use of smart structures to solve vibration suppression problems in situations where conventional sensors and actuators are not applicable. From a control theory point of view, the use of the smart structure approach is beneficial because it allows the control designer to measure the structure's response at many locations, rather than having to resort to estimation from a limited number of sensors. Similarly it is always better to provide the control force using a large number of actuators distributed throughout the structure. In addition, the use of smart structures allows placement of sensors and actuators at almost any location allowing a maximum use of the concepts of controllability and observability. From a mechanical design point of view, the use of smart structures offers an order of magnitude reduction in settling times for a small expenditure of power.

3.4.1 *A Cantilever Beam*

A programmable structure is defined as a structural element (for example a beam, plate or shell) that contains an integral closed loop system consisting of sensor, actuation and control logic with adjustable control parameters. The modular aspect of the programmable structure is beneficial to large or complex mechanical systems such as a truss which consists of many smaller structures. Similarly, it would be useful in forming a smart skin. The removal and replacement of one or more of the sections of a truss or smart skin would be facilitated by the use of modular components. Through the use of redundant programmable elements in a structure containing many programmable structural elements, loss of any single element will not necessarily lead to a failure in meeting the performance objective. To illustrate the concept of a smart structure used here, as well as to provide an introduction to basic elements used in the following examples, consider the programmable cantilever beam shown in Fig. 3.12. This beam consists of

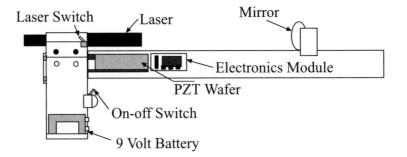

Fig. 3.12   Schematic of a programmable beam.

eight layers of fiberglass, and near the clamped end of the beam piezoceramic material is embedded in place of the fiberglass in the second and seventh layers, as indicated in Fig. 3.13. The beam is 479mm long, 33mm wide and has a thickness of 2mm.

A prototype programmable beam has been constructed to illustrate the capabilities of a smart structure. The programmable beam is a self-contained unit with sensors, actuators, power amplifier, and control circuitry imbedded in and mounted on the beam. The power supply, a common 9V battery, is attached to the base of the beam, and is used to emphasize the low power requirements. The fiberglass has a ply thickness that is equal to the thickness of the piezoceramic material used for actuation and sensing. Smooth integration of the ceramics into the fiberglass layers is therefore possible.

A flat, adhesive backed, copper lead is attached to each side of the piezoceramics. The leads transmit signals to and from the ceramic and are attached the entire length of the piezoceramic. In doing so, electrical contact will still be maintained even if the piece cracks in a direction normal to the direction of the lead. A hole is cut out of the second and seventh layers of the fiberglass ply which corresponds to the size of the piezoceramic. The piezoceramic with attached leads is placed in this hole and the remaining plies are layered to form the composite. Holes are not cut out of the fiber-

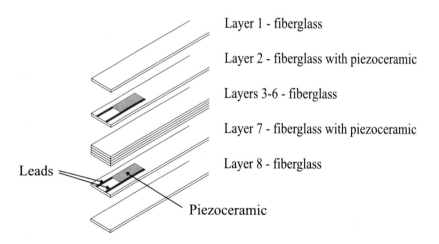

Fig. 3.13   Exploded view of the cross section of the beam illustrating the out of phase pair of embedded actuators.

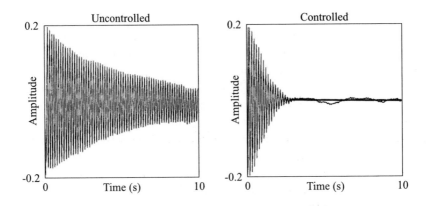

Fig. 3.14 Open and closed loop response of the cantilever beam with two PPF filters.

glass to accommodate the copper leads since they are much thinner than the plies.

The controller, surface mounted in this case, consists of the electronics required for self sensing actuation, as described in Sec. 3.2. The uncertainty in the parameters of a layered material demand that the closed loop control system be chosen to be robust to parameter changes, yet allow a substantial amount of damping to be added to the structure. Such a control scheme is provided by independent modal space control using positive position feedback, as described in Sec. 3.3.3. Figure 3.14 shows the success of the controller in producing an order of magnitude increase in damping. In fact the settling time for the tip response decreases from about 30s open loop to around 3s closed loop, at an energy cost of about 0.038W. The figure shows the response of the tip of the beam, as measured by a proximity probe, resulting from an initial tip displacement.

While this example provides experimental verification that the programmable structural concept works, there are alternative strategies for controlling cantilever beams. Similar performance can be obtained by attaching a constrained layer viscoelastic material to a portion of the beam. However this active cantilevered beam result is used in the following as a building block to solve more complicated vibration suppression problems. In the remaining sections piezoceramics, self sensing actuation and PPF control are used in successively more complex applications. These applications are presented to illustrate logical uses of the technology referred to as smart structures. In each case that follows, it is difficult to see how the same performance could be achieved without the use of devices such as piezoceramics in an integrated smart structure approach.

3.4.2 A Slewing Beam

Although controlling the vibration of a cantilever beam is relatively straightforward, when combined with the slewing motion of a flexible beam some very interesting phenomena occur. Slewing motion occurs when an object is swung around a fixed point, such as a door on a hinge. This brings to mind the slewing motion common to robotics, manufacturing and satellite applications (such as slewing solar arrays) and aircraft control surfaces. Slewing, as demonstrated here, involves the use of an electric motor as a rigid body actuator to reposition a beam in a new direction. Here the direction of slewing motion is also the bending direction of the beam as indicated in Fig. 3.15.

This section compares the response of a slewing flexible beam, both with and without a piezoceramic sensor–actuator pair mounted on the surface of the beam. Two important conclusions result. The first is that the active beam can produce the same performance as the slewing passive beam with

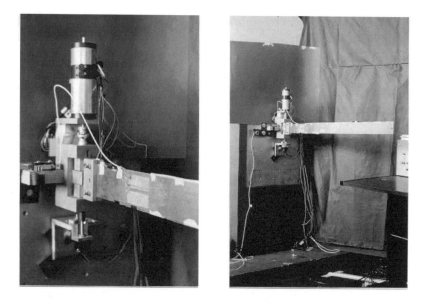

Fig. 3.15   Slewing beam with piezoceramic patches attached.

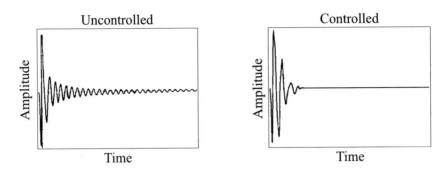

Fig. 3.16   Effect of the piezoelectric actuator control on the closed loop response of the slewing beam.

about 30% less power to the primary or rigid body actuator (the motor). The second result is that the active beam can be used to substantially improve the performance of a closed loop slewing maneuver.

The system under consideration is a slewing flexible structure, a thin aluminum beam, driven by an electric motor and actuated by a piece-wise distributed piezoceramic actuator. An improvement in performance is gained by modeling the motor and beam interaction and by using the piezo-electric actuators for direct vibration suppression of the beam dynamics. This presents a multi input slewing control problem which is implemented using a standard linear quadratic regulator control design, as described in Sec. 3.3.4, using a model estimated from the measured system dynamics. The active beam being considered here is similar to those considered by Fanson and Caughey [6].

The experimental results are illustrated in Fig. 3.16. The slew maneuver requires less control input when the piezoceramic actuator is used, and less power for the rigid body actuator to achieve the same relative performance, as measured by the settling time. Reduced control effort in the electric motor means a smaller motor could be used. On the other hand, if the best performance becomes the goal, Fig. 3.16 illustrates that the use of the secondary piezoceramic actuator improves the performance by almost an order of magnitude.

This result is not surprising if the issue of controllability is considered. In general a two input control system is more controllable than a single input control system. In fact the ideal situation is to use full state feedback which would require actuators for each mode that is important for the beam dynamics. While previously thought to be impractical, the use of piezoceramic actuators helps to approach the ideal situation of full state feedback. This thought is highlighted in the next section on slewing frames.

3.4.3   *A Slewing Frame*

Next consider complicating the dynamics a little more by considering a slewing frame, where the torsional flexibility of the frame becomes important. Here the smart structure solution becomes even more interesting as smart structures clearly improve the controllability of the closed loop system. The primary action of slewing induces both bending and torsional vibrations in the structure. Two active members that can be used as collocated sensor/actuators in feedback control loops are inserted into the frame. Control laws are designed that simultaneously slew the frame and suppress the residual vibrations. Results indicate that the motor alone is effective in slewing the frame and suppressing the bending motion but not the torsional motion. Hence, the torsional vibrations are suppressed using the active members. A complete description of the model and controller used is given by Leo and Inman [16].

The frame consists of individual elements of thin-walled circular aluminum tubing. Each member is 6.35mm in diameter and has wall thickness of 1.24mm. The elements are joined at octagonal nodes that are also made of aluminum. Each member is pinned and bolted at the node to eliminate looseness in the joints. The frame is mounted onto the larger steel shaft by bolting two of the nodes into aluminum clamps. A tachometer is mounted

Table 3.1   Natural frequencies and damping ratios for the slewing frame using the motor only and a PD controller.

Mode	Analytical		Experimental	
	Natural Freq. (Hz)	Damping Ratio (%)	Natural Freq. (Hz)	Damping Ratio (%)
Torsional	4.37	0.41	4.32	0.82
Bending	8.87	11.24	7.68	9.18
Torsional	15.47	0.50	14.11	1.26
Plate	19.79	0.53	20.76	0.94
Bending	27.53	5.10	26.25	1.32

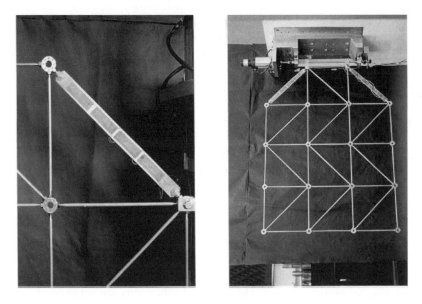

Fig. 3.17   Slewing frame test bed showing the location of the active members, angular rate and position sensors.

inside the motor measures angular rate, and a potentiometer attached to the bottom of the steel shaft produces a signal proportional to angular position. The whole slewing rig is attached to a large concrete block to provide a rigid support. Figure 3.17 shows pictures of the slewing frame test bed.

Several controllability measures are available. These measures can provide an indication of where to place actuators on a structure to achieve the best closed loop behavior. In particular a low controllability for a specific actuator location indicates the need for additional actuators. Both the analytical and experimental models of the closed loop response of a simple proportional–derivative (PD) control are listed in Table 3.1. The results show low damping in the torsional and plate modes as well as good agreement between the analytical model and the experimentally measured frequencies and damping ratios. The closed loop damping ratio is used here as a performance index. Note that the closed loop damping ratio for the two torsional modes and the plate modes are much lower than the damping ratios for the bending modes. This indicates that the motor control alone is not able to damp the torsional nodes and suggests a controllability problem. This is illustrated in Fig. 3.18 which shows the response at the active

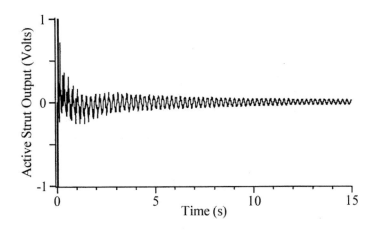

Fig. 3.18   Controlled response of the slewing frame at the active strut, using only the motor.

strut when the motor only is used for control. The lightly damped torsional mode is clearly visible in the response and has a very large settling time.

Because of the inability of the motor actuator by itself to control the torsional modes, the secondary piezoceramic actuators are added to the frame as illustrated in Fig. 3.18. Adding these secondary actuators improves the controllability as can be seen by examining any number of controllability measures. Table 3.2 illustrates the application of such a controllability measure to the two systems, namely the motor control of the frame and motor control plus a piezoceramic actuator. Tables 3.1 and 3.2 illustrate the difficulty in controlling such a system and how embedded piezoceramic

Table 3.2   Controllability of the modes of the slewing frame.

	Gross Controllability Measure	
Mode	Slewing actuator alone	With active member
1st Torsional	8.62	109.55
1st Bending	155.16	156.42
2nd Torsional	1.11	39.81
1st Plate	17.28	54.31
2nd Bending	80.29	91.61

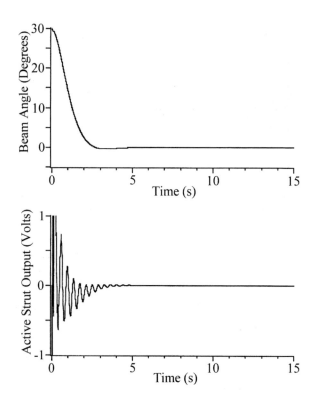

Fig. 3.19   Controlled response of the slewing frame using both the motor and the active strut.

actuators provide straightforward and logical solution. The motor causes motion in the torsional and plate modes of the frame, yet this actuator as a control device offers very little control authority over these modes. The addition of the embedded piezoceramics on the rotating structure renders these troublesome modes controllable. The result is a faster, more well behaved response, as illustrated in Fig. 3.19.

Here it is also important to note that the cost of this tremendous gain in performance is very little. In fact the total power increases from 121.53W to 121.84W, while the damping ratio increases by a factor of 15 in the first torsional mode, 2.3 in the second torsional mode and 6 in the first plate mode. Thus a very inexpensive addition in control complexity and effort results in a tremendous increase in performance providing an excellent example of the utility of the smart structures concept.

3.4.4 *Antenna*

Next theoretical and experimental results of the modeling and control of a smart ribbed antenna are presented. The antenna consists of eight flexible ribs which constitutes a smart antenna in the sense that the actuator and sensors are an integral part of the structure. The antenna has modes with closely natural frequencies and also modes that have the same natural frequency (repeated modes). Controllability and observability considerations mean that multi-input multi-output (MIMO) control is necessary. The nearly periodic nature of the structure, and shape of the ribs makes the structure difficult to model accurately using finite element analysis. The periodic structure of the antenna means that small perturbations in the manufacturing tolerance of parts of the structure can lead to large changes in the dynamics of the antenna [18]. An identified model of the antenna is synthesized from curve fitting the measured FRF data in the frequency domain [4]. The identified model is used to design a positive position feedback (PPF) controller that increases the damping of all of the modes in the targeted frequency range. Due to the accuracy of the open loop model of the antenna, the closed loop response predicted by the identified model correlates well with experimental results. This example gives a unique solution to a difficult class of structural control problems, namely those with a high modal density.

A schematic of the antenna is shown in Fig. 3.20. Located at the base of five of the test antenna's ribs are collocated piezoceramic sensor/actuator pairs. Each collocated sensor/actuator is manufactured from a single sheet of piezoelectric material. The electrode surface of the piezoceramic is separated into two electrically isolated areas. Each rib is fixed at the base to a rigid hub and connecting rib's tips are tensioning wires. During the manufacture of the antenna the ribs are initially flat and buckled into a dish shape by tension in the wire connected at the tip of each rib.

Typical open loop frequency responses for the frequencies from 9 to 19Hz are shown in Fig. 3.21. A cluster of seven closely spaced modes is observed between 9.6 and 17Hz. A second cluster of modes can be observed near 30Hz and subsequent clusters occur at higher frequencies. The first group of modes corresponds to each individual rib vibrating in their first mode, the second cluster corresponds to each rib vibrating in its second mode, and so on. There are eight ribs and it would be expected that eight modes exist in the first modal cluster. Only seven can be observed, and thus there is a repeated mode (two modes with the same natural frequency) in this first cluster. The existence of this repeated mode indicates that at least two actuators and two sensors are required for controllability and observability [13]. The identified model is valid over the frequency range from 0 to 20Hz that covers the first modal cluster.

From a strict definition of controllability, it is known that a minimum of two active ribs are required on the experimental antenna to control the repeated mode at 11.01Hz. A controllability measure is used to determine which two active ribs should be used to provide the greatest degree of controllability. Table 3.3 gives the controllability measures for different pairs of ribs. This is similar to the measure used in Table 3.2 for the slewing frame, except here the values are normalised to have a maximum value of 1. The table is based on the method of Hamden and Nayfeh [12] for the distinct modes and the method of Hughes and Skelton [13] for the repeated mode. The lowest value of controllability at each frequency for each choice of actuator location is highlighted in the table. Note that locations 1 and 5 give the largest low value (max–min). Thus the controllability measures with active ribs 1 and 5 are consistently high when compared the controllability measures calculated for other pairs of ribs. Using these active ribs will ensure the best controller performance for the lowest cost.

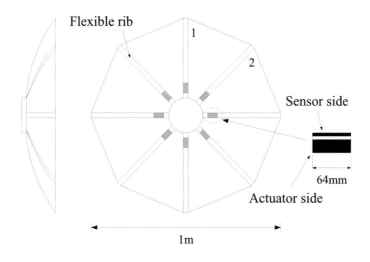

Fig. 3.20   Schematic of the active antenna.

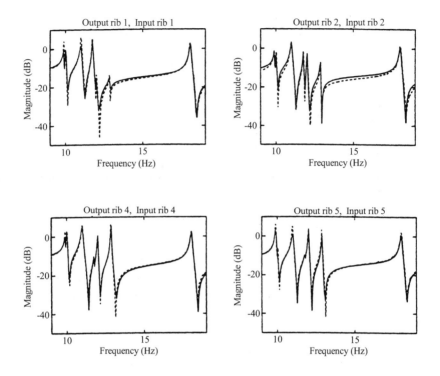

Fig. 3.21 Four typical open loop frequency response functions of the active antenna (solid is the predicted, dashed is experimental)

PPF control laws are implemented on the smart antenna and the closed loop response compared with the response predicted by the model. The PPF controller is not affected by unmodeled high frequency dynamics resulting in better experimental performance. In fact it can be shown that as the gain is increased with the PPF controller the instability will result at low frequency [6], and the low frequency dynamics are well modelled. A constrained optimization routine is used to design the PPF controller parameters. The optimization minimizes the weighted sum of the response of the antenna model to a unit impulse and the control effort. Values for the weight are adjusted to achieve the best trade off between minimizing the impulse response and minimizing the control effort. The control is optimized for the closed loop control operating on ribs 1 and 5. In the control design each actuator is capable of having multiple filters associated with it,

Table 3.3   Controllability measures for pairs of ribs on the smart antenna.

| Mode Freq. (Hz) | Controllability Measure | | | | |
	Ribs 1 & 2	Ribs 1 & 4	Ribs 1 & 5	Ribs 2 & 4	Ribs 2 & 5
18.0480	0.9102	0.9419	0.9067	0.9071	0.8705
12.8800	**0.3089**	0.5455	**0.4188**	0.5919	0.4777
12.0143	0.3662	0.4319	0.5034	0.5244	0.5848
11.7736	0.6661	0.5806	0.5770	**0.3822**	0.3766
10.0326	0.5493	0.6626	0.4829	0.6043	**0.3991**
9.9087	0.5372	0.5122	0.6944	0.4909	0.6788
11.0103	0.3678	**0.1654**	0.4379	0.3202	0.2640

and so more than one mode per actuator is considered. It is found that no significant reduction in the cost function is obtained using multiple filters. This can be attributed to the fact that the modes are closely spaced and the model does not include higher frequency modes. Therefore, only a single filter per actuator is used in the implemented design. If the response and control effort are equally weighted then the optimal filter parameters are a natural frequency of $70.2 \mathrm{rads}^{-1}$ and a damping ratio of 0.56.

Figure 3.22 compares the experimental response of the antenna with PPF control on ribs 1 and 5, with the model's predicted response and also with the open loop response. A significant increase in damping has been achieved in all the modes in the first modal cluster using only a single PPF filter on each of the two active ribs.

3.4.5   *Plate Example*

The purpose of this last application is to illustrate that smart structures can compete successfully and out perform traditional damping treatments in situations where the operating temperature varies. The system considered here uses a piezoceramic sensor and actuator along with a temperature sensor (a 10kW thermistor) to form a control system based on an adaptive positive position feedback control law. Adaptive control laws have the unique feature of adjusting control gains to suit the current values of a system's parameters [20] for a general discussion of adaptive control). Constrained layer and free layer damping treatments made from viscoelas-

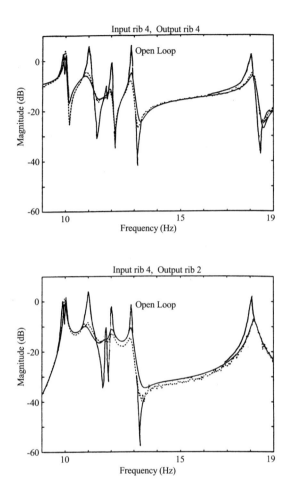

Fig. 3.22 Comparison of the open loop (labelled solid line) and closed loop responses (experimental dashed, predicted solid) of the antenna.

tic materials are commonly used to suppress vibrations in the floor panels of automobiles and fuselage panels in aircraft. One difficulty with using viscoelastic based treatments is that their damping properties fall off as the temperature increases, reducing the amount of vibration suppressed. The system presented here illustrates that smart materials may be used to provide active damping that retains significant vibration suppression while undergoing ambient temperature changes.

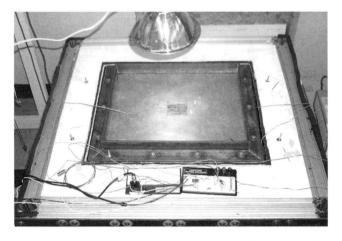

Fig. 3.23   Test rig for the vibration suppression of a plate in the presence of temperature changes.

The control system is applied using a standard test plate used on a standard test stand common to the automotive industry. The plate is a 20 gauge galvanized steel plate with dimensions of 500×600mm. To ensure repeatable boundary conditions the 14 clamping bolts were tightened in the same criss-crossing pattern to a torque of 25Nm. By clamping the plate in the excitation frame, the effective area of the plate is reduced to 400×500mm. Figure 3.23 shows the test plate mounted in the frame. The actuator patch used is a commercially available wafer of dimensions $72.4 \times 72.4 \times 0.267$mm. Because the objective of this research was to increase the damping of the first three to four vibration modes, the PZT actuator was placed in the center of the plate. The temperature was varied by using a 250W heat lamp, placed about 300mm above the plate and this was able to heat the center of the plate up to 70°C. Due to the location of the heat lamp over the centre of the plate the distribution of the heat was not very uniform. While the highest temperature was always reached in the centre, the temperature decreased significantly towards the boundaries of the plate.

A primitive form of adaptive control was implemented by designing a set of PPF filters for each 2.5°C temperature increment. As the temperature changed to a new value the natural frequencies of the clamped plate changed drastically as illustrated in Fig. 3.24. At each temperature the three filter PPF gains were calculated in advance. The control law then changed its gains according to the temperature at the centre of the plate.

The results are illustrated in Fig. 3.24. Note that the first three modes are still suppressed (have low peak values) even though the temperature has risen.

Viscoelastic materials are in general not able to provide damping across this range of temperatures and are usually designed at a fixed temperature.

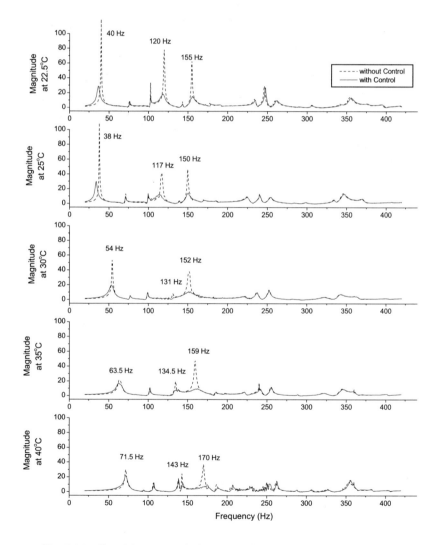

Fig. 3.24 Closed loop control of the plate for five different temperatures.

Using active control with some sort of adaptive feature allows the vibration of the plate to be suppressed even in the presence of large temperature swings.

3.5 Conclusions

This chapter has introduced some of the methods to reduce the vibration in flexible lightweight structures using smart materials. The concepts required for the dynamic analysis of structures have been introduced to enable the control strategies and performance to be placed in context. The examples have illustrated the use of smart structures in a logical way to solve control problems of increased complexity. The smart structures philosophy provides an inexpensive means of introducing feedback control of structures, without adding significant mass or requiring significant power. Smart actuators provide a means to control modes of structures that are not easily controlled by any other means.

Bibliography

[1] Avitabile, P. (1998) "Modal Space: Back to Basics", series of articles in *Experimental Techniques*, also available at www.eng.uml.edu/macl/umlspace/mspace.

[2] Dosch, J. J., Inman, D. J. and Garcia, E. (1992) "A Self-Sensing Piezoelectric Actuator for Collocated Control", *J. Intel. Mat. Sys. & Struct.*, **3**, 166–185.

[3] Dosch, J., Leo, D. J. and Inman, D. J. (1993) "Modeling and Control for Vibration Suppression of a Flexible Smart Structures", *Dynamics and Control of Structures in Space II* (ed. C. L. Kirk and P. C. Hughes), pp. 603–618.

[4] Dosch, J. J., Leo, D. J. and Inman, D. J. (1995). Modeling and Control for Vibration Suppression of a Flexible Active Structure, *J. Guidance, Control & Dynamics*, **18**, 340–346.

[5] Ewins, D. J. (2000) *Modal Testing*, Research Studies Press.

[6] Fanson, J. L. and Caughey, T. K. (1990) "Positive Position Feedback Control for Large Space Structures", *AIAA Journal*, **28**, 717–724.

[7] Friswell, M. I. (2001) "On the Design of Modal Actuators and Sensors", *J. Sound & Vib.*, **241**, 361–372.

[8] Garcia, E. and Inman, D. J. (1990) "Advantage of Slewing an Active Structure", *J. Intel. Mat. Sys. & Struct.*, **1**, 261–272.

[9] Garcia, E., Dosch, J. J. and Inman, D. J. (1991) "The Application of Smart Structures to the Vibration Suppression Problem", *Proc. 2nd Joint Japan–US Conf. on Adaptive Structures*.

[10] Garcia, E., Dosch, J. and Inman, D. J. (1992) "The Application of Smart Structures to the Vibration Suppression Problem", *J. Intel. Mat. Sys. & Struct.*, **3**, 659–667.

[11] Goh, C. L. and Caughey, T. K. (1985) "On the Stability Problem Caused by Finite Actuator Dynamics in the Collocated Control of Large Space Structures", *Int. J. Control*, **41**, 787–802.

[12]   Hamden, A. M. A. and Nayfeh, A. H. (1989) "Measures of Modal Control-lability and Observability for First- and Second-Order Linear Systems", *J. Guidance, Control & Dynamics*, **12**, 421–428.

[13]   Hughes, P. C. and Skelton, R. E. (1980) "Controllability and Observability of Linear Matrix Second Order Systems", *J. Appl. Mech.*, **47**, 415–420.

[14]   Inman, D. J. (1989) *Vibration with Control, Measurement and Stability*, Prentice Hall.

[15]   Inman, D. J. (1994) *Engineering Vibration*, Prentice Hall.

[16]   Leo, D. J. and Inman, D. J. (1993) "Modeling and Control of a Slewing Frame Containing Self-Sensing Active Members", *Smart Mat. & Struct.*, **2**, 82–95.

[17]   Leo, D. J. and Inman, D. J. (1994) "Pointing Control and Vibration Suppression of a Slewing Flexible Frame", *J. Guidance & Control*, **17**, 529–536.

[18]   Levin–West, M. B. and Salama, M. A. (1992) "Mode Localization Experiments on a Ribbed Antenna", *Proc. 33rd AIAA Structures, Structural Dynamics and Materials Conf.*, pp. 2038–2047.

[19]   Meirovitch, L. and Baruh, H. (1982) "Control of Self-Adjoint Distributed Parameter Systems", *J. Guidance & Control*, **5**, 59–66.

[20]   Slotine, J. –J. E. and Li, W. (1991) *Applied Nonlinear Control*, Prentice Hall.

Chapter 4

Data Fusion — The Role of Signal Processing for Smart Structures and Systems

Keith Worden and Wieslaw J. Staszewski
Dynamics Research Group,
Department of Mechanical Engineering,
University of Sheffield,
Mappin Street, Sheffield S1 3JD, UK.

4.1 Introduction

The ideal *smart structure* is one which is capable of sensing its environment and is also aware of its current state of condition and motion. It is capable of diagnosing damage to itself and taking action to minimise the consequences. It is capable of adapting its static and dynamic behaviour through feedback control. A prerequisite for such a structure is that it should have *integrated* sensing and actuation capabilities. That is the capability for sensing and movement should be built-in, but perhaps more importantly — designed-in.

Many types of sensor are available, some localised, some distributed. Each has its own advantages and disadvantages. The generic smart structure will need to process information from sensors of both types. Remote sensing is necessary for identifying possible hazards and harsh environments and providing information to the control algorithms which will direct the structure to avoid problems. Distributed sensing may well be needed for the location and assessment of structural damage.

The efficient use of integrated sensors of different types requires the availability of appropriate signal processing technology. Sophisticated *pattern recognition* algorithms are needed. In the case of remote sensors, data representing possible hazards must be classified so that appropriate action can be taken. In many military applications, target recognition is critical. This is the concern of *image processing*, the subject of much current research. In the case of distributed sensors for damage detection say, the vector of data returned must be translated into a diagnosis of location and severity. This is also a field of study which has seen intense activity lately which has proved fertile. In the situations where sensors of different types return information, *data fusion* techniques are precisely what is needed which extract meaning from disparate sources. This chapter attempts to give a unified account of a number of fusion strategies.

In terms of *intelligence* the smart system will require a central processing unit or distributed processing network, which not only carries out data fusion from the available sensors, but decides action on the basis of the results. The relevant disciplines here are control engineering and artificial intelligence. Animals are capable of learning from experience and this is clearly a desirable property for a smart structure. Artificial neural networks (ANNs) offer one of the most promising means of implementing this mechanism. Neural networks are also extremely effective pattern recognition systems and may well play a significant role in managing and interpreting sensors. Although the detailed structure and function of neural networks is not covered in this chapter, the reader may refer to the Appendix for a description of one of the most common types.

The word smart is also applied to certain classes of materials. Although a material cannot be truly smart under the definitions above, it can have properties such as adaptability which make it a desirable element of smart structures. Computers have an important role to play in the essentially passive activity of modelling and simulation of materials.

This chapter is not intended to be a survey of smart technology in the broad sense, the physics of the various sensors discussed will largely be ignored as the reader will encounter this material in the other chapters here. Rather, the object of this chapter is to overview applications of data fusion to smart structures and materials. This will be accomplished by giving the relevant theory for a number of common fusion methods and also reviewing two recent applications.

Of the various requirements for a smart structure or system discussed above, one — that of remote sensing and target and track identification — has already accumulated a large body of literature and would require a specialist study in its own right. The reader is referred to [1] for a tutorial description of the relevant matters and also for initial pointers into the literature. Only a scattering of results will be reported here where the methods are of particular interest. The current survey will mainly concern itself with issues which are applicable to military and civil systems in equal measure. These issues will be sensor fusion for *control* and for *damage assessment* in smart systems.

Given the priorities for smart structures, it may be more appropriate to speak in terms of *sensor–actuator fusion*. It is noted in [1] that one of the simpler fusion systems could simply pass control from a wide-range low-resolution sensor to a narrow-range high-resolution sensor in order to refine the search-and-identify task. The low-resolution sensor would need to communicate area and range for search to the high-resolution unit. A related problem can arise in smart systems where a sensor must pass details of condition or environment on to an actuator in order that the system can make a control action.

Anyway, before any talk of sensor fusion, it is important to have an idea of what a sensor is and what a sensor does. This is the subject of the next section.

4.2 Sensors

As discussed above, the term *sensor* can be interpreted broadly as a provider of data. In the military framework, vital information may be provided by a variety of channels and it is there that the interpretation is broadest. For the purposes of this review, a more restricted definition of sensor may be adopted without loss of generality. A sensor will be understood to be some physical construct which translates an environmental or structural variable into an electrical signal which can be quantified. (There is little point in attempting further precision as this may lead to discussions bordering on philosophy.) In most structural dynamic applications, the sensor signals are processed — either in software or hardware — and used to formulate some decision about the structure. This may be a control decision, where it proves necessary to alter the state of the system, or it may be a health management decision where the system is taken off-line for more detailed inspection or repair.

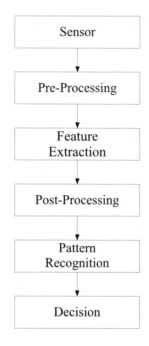

Fig. 4.1   Processing chain for a single sensor.

The processing chain for an individual sensor is therefore as summarised in Fig. 4.1. The elements of the chain can be summarised as follows:

Sensor Provides an electrical* signal proportional to the structural or environmental variable of interest. For example, an accelerometer could be placed on the outer casing of a gearbox in order to monitor its condition or state of health.

Pre-Processing In most circumstances, the time-varying voltage from the sensor will contain too much data, much of which is redundant. The goal of the pre-processing stage is to reduce the dimension of the data vector and eliminate as much redundancy as possible. An example of this step for the gearbox would be to convert the time data to a spectrum, i.e. a description of the dominant frequencies in the signal.

---

*This could be acoustic or thermal in all generality. However, it is convenient to assume that the sensor is interfaced to an acquisition/interpretation system that requires a charge or voltage.

The windowing and averaging process for this step would result in far fewer spectral lines than the points of the original time-series. This step is usually carried out on the basis of experience and engineering judgement. At this stage the aim would be to reduce the dimension of the data vector from possibly many thousands to about a hundred.

Feature Extraction Again, the object is dimension reduction and elimination of redundancy. This is considered a separate step as the techniques will usually be based on statistics or information theory and not on Engineering judgement. For example Principal component analysis (PCA) could be used. The aim of this stage would be a further reduction in dimension to less than ten channels of information-rich data or *features*. (A low-dimensional feature set is a critical element in any pattern recognition problem as the number of data examples needed for building a diagnostic grows explosively with the dimension of the feature space.) An example for the gearbox problem which does in this case use Engineering intuition would be to select lines from the spectrum at the meshing frequency[†] of the gearbox and at higher harmonics. It is also known that significant information is contained in the sidebands of the harmonics.

Post-Processing This is simply preparation for the pattern recognition stage and may amount to no more than selection of an overall scaling or *normalisation* for the feature vectors consistent with the pattern recognition algorithm, e.g. some neural networks require all the data to be between -1 and $+1$. Alternatively, more advanced processing is possible; for example the the feature data may be nonlinearly transformed in order to produce data with a Gaussian distribution.

Pattern Recognition This is the critical stage. The feature vectors are passed to an algorithm which can classify the system on the basis of the measurements. An example here would be a neural network which is trained to return a fault-type and severity when presented with the restricted spectral data from the gearbox.

Decision The final stage — which may be under computer control or require human intervention — is a decision and action on the basis of the pattern diagnosis. To conclude the gearbox example, this would entail a decision on whether to shut the system down or not depending on the severity of the fault.

[†]The frequency at which the teeth on two meshing gears come together.

One thing which is immediately obvious from Fig. 4.1 is that a single sensor strategy is rather fragile. If the chain breaks at any point, the decision making facility is lost. This is one of the compelling reasons for moving to a network of sensors wherever possible.

4.3   Sensor Fusion

Sensor and data fusion as a research discipline in its own right, emerged largely as a result of various Defence organisations attempting to formalise procedures for integrating information from disparate sources. The object of the exercise was to determine battlefield situation and assess threat on the basis of data coming in from numerous different channels. A *sensor* in this context is simply a provider of data and may be a true sensor in the physical sense like a microwave antenna or even a human intermediary. The *philosophy* of data fusion was quickly recognised to have broader application; initially in the fields of meteorology and traffic management, but later in the medical field and in non-destructive testing (NDT).

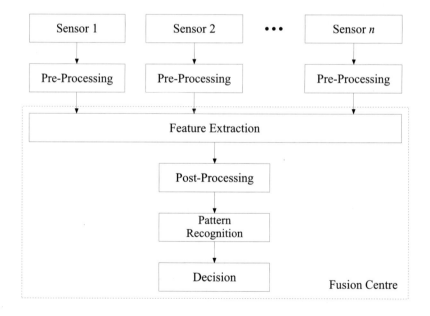

Fig. 4.2   Centralised fusion.

Thus, the whole purpose of *sensor fusion* as a discipline is to integrate data from a multitude of sensors with the objective of making a more robust and confident decision than is possible with any one sensor alone. There are numerous reasons why multi-sensor systems are desirable, following [2]:

- Higher signal-to-noise ratio. (Sensor fusion can be as simple as averaging the results from several identical sensors. In practice, ergodicity[‡] helps here by allowing averages over sensors to be replaced by averaging over time.)
- Robustness and reliability. Enough information may be available to form a decision even if a subset of the sensors fail. Note that in order to have fault tolerance, it is necessary to design in redundancy and thus increase the dimension of the measurement. This redundancy should always be removable in the signal processing or the dimension of the pattern recognition and decision problem will increase.
- Information regarding independent features in the system can be obtained. This is clearly desirable; for control problems, all the states of interest should be observable from the measurement set; for health monitoring, the measured variables should be sensitive to all aspects of the damage.
- Extended parameter coverage gives a more complete picture of the system.
- Improve resolution (averaging can reduce quantisation error).
- Increased confidence in the results. This may amount to little more than agreement by several independent measurements, or could go as far as providing statistics for uncertainty.
- Increased hypothesis discrimination.
- Reduced measurement times (under certain circumstances).

There are many ways of implementing fusion strategies. Common to all is that the single sensor processing chain of Fig. 4.1 is replicated a number of times and the chains are fused together. An example strategy, which is usually called *central-level* or *centralised* fusion [1] is shown in Fig. 4.2.

Each of the sensors provides information to a central feature extraction unit which thus attempts to remove both *inter-* and *intra-*sensor redundancy (between sensors and within individual sensors). Another approach

---

[‡]Loosely speaking, the property that the overall characteristics of the signal do not change over time. It is actually a stronger property than this.

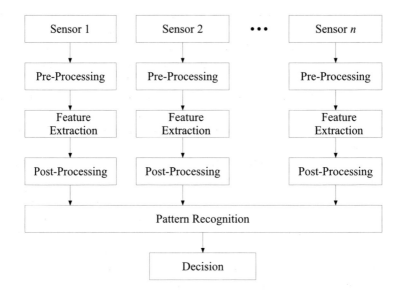

Fig. 4.3   Pattern-level fusion architecture.

could be a pattern-level fusion architecture as in Fig. 4.3, where the feature extraction is carried out for each sensor independently.

While there are broad categories to define here, one cannot help but feel that it is largely unprofitable to classify fusion *architectures* in the manner above as each situation could call for a different fusion strategy. There is no reason why individual sensors chains should not be fused at different points and this means that there are as many fusion strategies as there are ways of connecting the chains (Fig. 4.4).

An architecture can therefore be thought of as a directed graph with input nodes the sensors and output nodes the decisions. The information flowing through the graph is initiated as the sensor values together with estimates of the sensor confidence and is condensed and refined at each stage of its passage and therefore continually changes its nature. Fusion occurs at the vertices and will require different techniques depending on the position or level of the vertex. The different fusion levels are as follows:

Pixel or Raw Sensor-Level Fusion The data from two or more of the sensors is combined at the pixel level or the level of the individual resolution cells. This occurs before any pre-processing is performed.

Feature-Level Fusion Two or more pre-processed sensor signals are combined to produce a single feature-vector for classification. This may be as simple as concatenation or may be as complex as a nonlinear mapping like nonlinear principal component analysis.

Pattern-Level Fusion Two or more feature-vectors are combined and passed to the pattern recognition algorithm.

Decision-Level Fusion Two or more decisions or classifications are combined in such a way as to produce a decision with higher confidence. Dempster–Shafer combination or Bayesian methods can be used here [1] (Fig. 4.5).

Most of the discussion so far has concentrated on taxonomy. While it is important to establish a common vocabulary, it is arguably more profitable

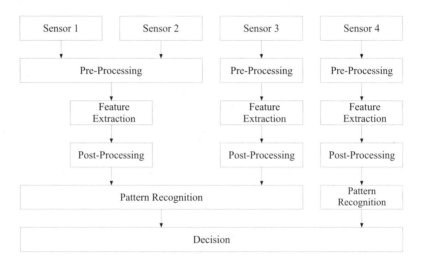

Fig. 4.4   Arbitrarily connected 4-sensor fusion tree.

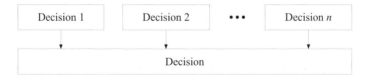

Fig. 4.5   Decision-level fusion.

to consider some of the standard models for fusion to see how they function and learn from them. The term *model* here is understood to mean a holistic framework which is centred on a fusion architecture but may also specify courses of action which follow from decisions returned by the fusion process. It may also allow for the possibility of feedback — modification of the architecture on the basis of decisions. This definition of a model is broadly in agreement with that in [3].

4.4   The JDL Model

The first attempts at formalising the discipline of data fusion were made in the military domain of *command, control, communications and intelligence* or C^3I. These efforts date back to the late seventies and early eighties. The early papers [4; 5], refer to *hierarchical structures, fusion system architectures* and *adaptive hypothesis testing*, all integral parts of modern fusion systems. A major step forward was made in 1990 when the North American Joint Directors of Laboratories (JDL) were charged, through their Data Fusion Subpanel (DFS), to establish a standard data fusion terminology and formulate a coherent strategy for exchanging technical information relating to fusion [6; 7]. The standard was augmented in [8] by making detection one of the objectives of fusion. The definition of data fusion arrived at was:

> "A multilevel, multifaceted process dealing with the automatic detection, association, correlation, estimation and combination of data from single and multiple sources."

The structure of the proposed data fusion model is summarised diagrammatically in Fig. 4.6. It was a hierarchical system of four (or five including pre-processing) levels as follows [1]:

Level 1 Achieve refined position and identity estimates by fusing individual sensor position and identity estimates.

Level 2 Assist in complete and timely hostile or friendly military situation assessment.

Level 3 Assist in complete and timely force threat assessment.

Level 4 Achieve improved results by continuously refining estimates and assessments, evaluating the need for additional sources, or modifying the process itself.

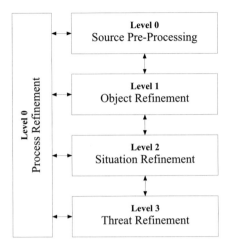

Fig. 4.6   The JDL data fusion model [16].

The description is couched entirely in military terms. However, note the similarity with Rytter's hierarchical structure for a damage monitoring system [9] which distinguishes four levels of damage identification:

Level 1 (DETECTION) The method gives a qualitative indication that damage might be present in the structure.

Level 2 (LOCALIZATION) The method also gives information about the probable position of the damage.

Level 3 (ASSESSMENT) The method gives an estimate of the extent of the damage.

Level 4 (CONSEQUENCE) The method offers information about the safety of the structure, e.g. estimates a residual life.

Rytter's level 1 corresponds to level 1 in the JDL fusion model, i.e. detection of whether a threat to safety is present. Rytter's levels 2 and 3 are condensed into level 2 of the JDL model, i.e. location and sizing of the damage is equivalent to assessing the situation. Finally Rytter's level 4 is broadly similar to the JDL level 3 in that it seeks to identify the likely consequences of the present danger. There are a number of observations one can make on this correspondence.

First of all, there is no analogue of the JDL level 4 in Rytter's scheme. This may be regarded as an oversight. However, Rytter's intention in forming the structure was simply to order and enumerate the different levels of information that should be available from a health monitoring system. It was not his purpose to define a holistic approach to damage identification. However, inspired by the precepts of the JDL model, one might go forward to defining such a holistic framework, which would include for example safety assessment of sensors, replacement and redeployment of sensors. These concepts are implicit in the JDL framework.

Secondly, in formulating the level 4 strategy for the JDL model, there is an implicit assumption that the initial conditions for the fusion system have been chosen in a careful if not optimal manner. It is well known amongst structural dynamicists that careful placement of sensors and actuators is a critical issue in a number of dynamical problems. The question of optimal placement will be illustrated later.

Thirdly; the 'sensors' referred to in the JDL model may be human or for that matter intelligent software agents [10]. On a smart structure, they will usually be physical devices. This means that questions of redeployment, etc., must be answered by mechanical means and this is potentially troublesome. The solution to this problem is to have redundancy in the sensor network, but this has cost implications. Replacement of sensors might possibly be avoided by the use of self-validating sensors [11; 12; 13] or fail-safe sensor networks [14].

4.5   The Boyd Model

The Boyd loop was initially used to model military command processes and is, like the JDL model couched in military terms [15]. However, like the JDL model, its domain of applicability is wider.

The Boyd model is commonly referred to as the *OODA model* or *loop* for obvious reasons. The four levels of the model are denoted *observe, orient, decide* and *act*:

Observe The sensor signals are accumulated. This is equivalent to JDL level 0.

Orient The accumulated information is used to form a picture of the situation and assess potential sources of threat. This level encompasses the three JDL levels 1, 2 and 3. It is the level at which sensor signals are fused and this also includes the incorporation of intelligence information.

Decision The possible courses of action are assessed and an optimal selection is made. These actions will include redistribution of sensors, etc., and this level therefore incorporates JDL level 4. However, it also considers actions external to the loop which make contact with the world outside the fusion model.

Act The plan decided in the decision phase is enacted. (There is no direct analogue of this phase in the JDL model and as a result the Boyd model is more general.) The consequences of the action on the environment should be assessed, and this forces a return to the observation phase. This closes the cycle.

In practice, the four phases of the loop are enacted in parallel and continuously. Data is accumulated continuously and used to refine the state of situation awareness; decisions and plans evolve throughout and are continuously brought into action.

It is observed in [16] that in a major operation, several OODA loops may be evolving in an interconnected manner. The reference in question also provides a description of a real application of the loop to naval warfare.

By including the possibility of action, the Boyd model provides a more suitable framework for discussing sensor fusion applications to smart structures. In control applications, the decisions made on the basis of sensor measurements will often have an impact outside the sensor chains. A simple example would be the use of shape-memory alloys to change an aircraft

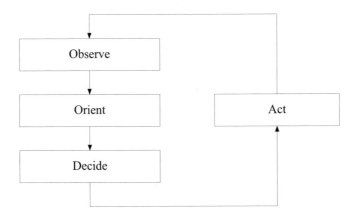

Fig. 4.7 The Boyd model.

wing profile in response to environmental measurements. For an essentially passive application like health monitoring the JDL would provide a sufficient framework; however, if active elements like self-repair are considered, the Boyd model is more appropriate.

4.6   The Waterfall Model

This is a model proposed in [17] which "has been widely used in the UK defence data fusion community but has not been significantly adopted elsewhere" [3]. The basic structure is shown in Fig. 4.8.

The model superficially appears to have many elements in common with the single-sensor processing chain of Fig. 4.1. However, this is an artefact of the simple diagrammatic representation. The sensing level in the waterfall model spans several sensors. Also note the *situation assessment* stage, this brings the model roughly into correspondence with the first three levels

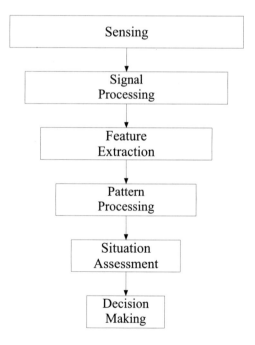

Fig. 4.8   The waterfall model.

of the JDL model. As observed in [3], the model lacks an explicit use of feedback corresponding to level 4 of the JDL model and is thus less general. The reason for discussing it here is that it provided one of the main ingredients for the Omnibus model described next.

4.7 The Omnibus Model

The reference [3] surveyed the fusion models described above as well as some others and identified that each has limitations. In an effort to design a model which incorporated all the desirable features of the standard fusion models and would overcome all their limitations, the authors drew up the following wish list. The ideal fusion model:

- defines the order of processing;
- makes the cyclic nature of the system explicit;
- admits representation from multiple viewpoints;
- identifies the advantages and limitations of various fusion approaches;
- facilitates the clarification of task-level measures of performance and system-level measures of effectiveness;
- uses a general terminology which is widely accessible;
- does not assume that applications are defence oriented.

The solution proposed in [3] is the *Omnibus model*. The model uses the fine levels of definition of the Waterfall model inside a cyclic structure reminiscent of the OODA loop. The structure is shown diagrammatically in Fig. 4.9.

The fact that this provides a unification of the previously discussed models is manifestly clear from the diagram. Because the Omnibus model is freed from the defence orientation of previous models and does not use a terminology specific to military applications; because it incorporates the best features of the influential JDL, OODA and Waterfall models, it is suggested that it is the most appropriate (currently available) framework for discussing applications to smart structures. Because of the explicit inclusion of the *control* level, the model actually provides an instance of sensor/actuator fusion as discussed earlier.

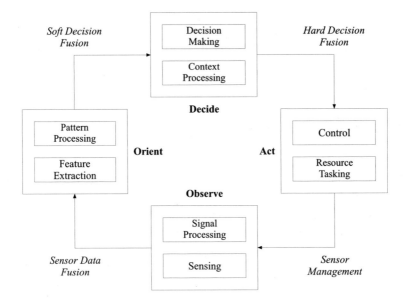

Fig. 4.9   The Omnibus model.

4.8   The Relevance of Data Fusion for Smart Structures

Previous sections have discussed some of the prevalent models and architectures for sensor/data fusion. The question is: what are the implications for the design and operation of smart structures?

The first lesson to be learnt from the fusion community is that one should as far as possible take a holistic view of the problem under consideration. A smart system should be designed with all aspects of its final operation under consideration.

- The most appropriate sensors should be selected. If necessary, a heterogeneous set should be chosen to cover different physical manifestations of the effects under investigation.
- The sensor distribution should be optimal. This means that the smallest number giving the required resolution should be active at any given time. Note that the optimisation may be constrained; there may be upper limits on the weight or power consumption of a monitoring system. If failure-safety is required, the distribution should if possible,

include inactive units which can be switched in in the event of sensor failure. This implies that the sensors themselves are monitored. This can be accomplished by using self-validating sensors or sensor networks. Active and inactive units should be at the optimal spatial locations.

- The fusion graph should be designed to give the most robust and confident decision process given the available sensor resolutions and confidences. Where possible, algorithms which are known to be optimal should be employed.

- If control is part of the system requirements, the actuator distribution should be optimal. The smallest number giving the required control actions should be active. As with sensors, redundancy can be used to give fault tolerance. The actuators should also be located optimally.

- Feedback should be an important part of the process. If the achieved decision process identifies an information gap and the system is capable of improvement by sensor/actuator redeployment, this should be implemented by whatever means available. This may be as simple as switching in inactive units or could in principle involve physical movement of the units.

- There should be a strategy for planning and implementing repair or replacement of the system. This may or may not allow the possibility of a back-up system, depending on cost constraints.

- Monitoring should be as continuous as possible, with all levels of the system updated as frequent as cost/technology constraints allow.

Now, one might argue that these considerations are largely common sense, and in a way this is precisely what sensor/data fusion models are designed to provide. They are intended as a means of formalising best practice for as wide a variety of data-driven problems as possible. In a sense they are analogous to the object-oriented paradigms for programming which contrast with the traditional problem-specific approaches to program design in that they force an organising framework on the programmer.

The second thing that data fusion as a discipline contributes is that it is a rallying point for all the algorithms which can be used to combine information at all its various stages of refinement. These are the algorithms which are applied at the vertices in the fusion graph. The later sections of this chapter will survey some of the more common techniques in some detail.

As the discipline of fusion has matured over the recent years, commercial software has become available which is applicable to multi-sensor fusion. Reference [18] provides a comprehensive survey of available software up to 1993. Although this paper will undoubtably contain references to obsolete products, it is a useful guide to the vendors in the field.

4.9   Case Study: Fault Detection Based on Lamb Wave Scattering

4.9.1   *Lamb Waves*

As described above, Rytter in [9] distinguishes four levels of damage identification:

Level 1 (DETECTION) The method gives a qualitative indication that damage might be present in the structure.

Level 2 (LOCALIZATION) The method also gives information about the probable position of the damage.

Level 3 (ASSESSMENT) The method gives an estimate of the extent of the damage.

Level 4 (CONSEQUENCE) The method offers information about the safety of the structure, e.g. estimates a residual life.

The first case study presents a method for establishing a reliable level one diagnostic. The technique, *novelty detection*, is essentially an extension of classical condition monitoring where one continuously observes a system parameter and signals any significant change. The method differs in that new signal processing techniques allow the synthesis of effective parameters from almost any measured pattern.

The problem considered here is delamination detection in a composite plate and the data used to define the novelty measure are time records of the intensity of waves, *Lamb waves*[§], travelling in the plate. The standard reference on the physics of Lamb waves is [19]. Lamb wave testing of composite plates has been discussed by numerous authors [20; 21 22; 23; 24; 25], describing the detection of various material failures including delaminations, fibre fracture and matrix cracking. One practical complication of Lamb wave inspection lies in the propagation characteristics of the elastic

---

[§]Essentially a type of wave which propagates only in *thin* plates.

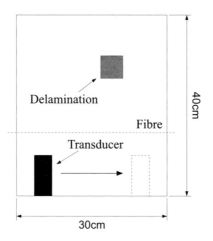

Fig. 4.10   Composite plate under test.

waves. The propagation is typically characterised by the product of the wave frequency and the sample thickness, or *frequency–thickness product* (*FT*). For low values of this parameter (typically, $FT < 1$MHzmm in aluminium), only two modes can propagate, the fundamental symmetric (S_0) and the fundamental antisymmetric (A_0).

The Lamb waves are launched with a conventional piezoceramic transducer and the resulting wave intensity (or amplitude) is recorded using an optical fibre interferometer. The arrangement of the test plate, and the location of the defect can be seen in Fig. 4.10. The delamination was simulated by placing a thin layer of PTFE material between two layers of the composite. (Considerably more detail regarding the experiment can be found in [26].) The use of optical fibre interferometers for the monitoring of Lamb waves has been investigated in depth by [25]. By subjecting the signal arm of an interferometer to the ultrasonic field, the stress-induced modulation of the refractive index of the optical fibre, leads to intensity fluctuation at the interferometer output. A sensitive photo-receiver thus allows observation of the acoustic wavefield.

The information about defects in the plate is encoded in the waves scattered by the fault. Because the time-variation of the wave intensity at the fibre-optic is quite complex due to the presence of boundaries, low reflection coefficients from the fault and the uncertain coupling between the transducer and the plate, it is advantageous to use an automatic pattern recognition technique to signal anomalous (faulted) condition. Clearly Lamb wave testing is considerably simplified if a known single mode is launched and this is the approach taken here, only the S_0 mode is allowed to propagate.

4.9.2   *Novelty Detection*

Neural networks have proved to be extremely powerful tools for pattern recognition [27] and they are adopted here for the fault diagnosis. A detailed description of the multi-layer perceptron (MLP) architecture used here is given in Appendix A for the curious reader. Briefly, artificial neural networks (ANNs) are vastly simplified models of the processing function of the brain. There are various structures in common use, but the MLP is arguably the most often used. This type of network is of *feedforward* type, i.e. signals propagate through the network in one direction only from the inputs to the outputs. The networks can be trained to reproduce any desired mapping capability. They can be used to form regression models, i.e. to predict a certain quantity on the basis of other measurements, and they can also be trained to classify, i.e. to assign a measurement to a given set of outcomes.

Note that if detailed models of the plate and defect were available, one could conceivably simulate the Lamb wave patterns corresponding to normal condition and various faults. If such *training data* were available, a neural network could be trained to provide a fault classification. This option was not available here due to the complex nature of Lamb wave interactions with delaminations in composite plates. Instead, a method was adopted based on the idea of *novelty* or *anomaly detection*; the network is trained only on normal condition data and is required to signal if there is a deviation from this condition. Such techniques can not provide classification information. The advantage is that the normal condition data can be taken from experiment and so the need for a detailed *a priori* model is removed. Examples of the use of novelty detection in medical diagnosis can be found in [28], case studies for engineering fault detection can be found in [29; 30].

The application to on-line damage detection is clear. It is assumed that damage will alter the measured patterns, so novelty will indicate a fault. The important point is to identify *significant* changes, i.e. those which can not be attributed to fluctuations in the measured patterns due to noise.

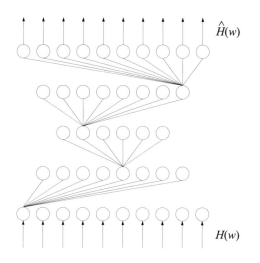

Fig. 4.11   Auto-associative neural network.

The approach taken here, based on [31], is simply to train an auto-associative network (AAN) on the patterns. This means a standard feed-forward multi-layer perceptron (MLP) network [32] which is asked to reproduce at the output layer, those patterns which are presented at the input. This would be a trivial exercise except that the network structure has a 'bottleneck', i.e. the patterns are passed through hidden layers which have fewer nodes than the input layer (Fig. 4.11). This forces the network to learn the significant features of the patterns; the activations of the smallest, central layer, correspond to a compressed representation of the input. Training proceeds by presenting the network with many versions of the pattern corresponding to normal condition corrupted by noise and requiring a copy at the output.

The novelty index $\nu(\underline{z})$ corresponding to a pattern vector $\underline{z} = z_i$, for $i = 1, \ldots, N$, is then defined as the Euclidean distance between the pattern $\underline{z}$ and the result of presenting it to the network $\hat{\underline{z}}$,

$$\nu(\underline{z}) = ||\underline{z} - \hat{\underline{z}}|| \tag{4.1}$$

It is clear how this works. If learning has been successful, then $\underline{z} = \hat{\underline{z}}$ for all data in the training set so $\nu(\underline{z}) \approx 0$ if $\underline{z}$ represents normal condition. If $\underline{z}$ corresponds to damage, $\nu(\underline{z})$ is non-zero. Note that there is no guarantee

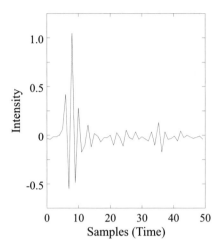

Fig. 4.12   Typical normal condition wave.

that ν will increase monotonically with the level of damage. This is why novelty detection only gives a yes/no diagnostic.

4.9.3   Results

The first stage of producing the novelty detector is to establish the pattern or set of measurements used for the diagnosis. In order to obtain the simplest possible diagnostic the raw time data were used with one or two modifications. Figure 4.12 shows the 500-point Lamb wave record for the leftmost source location. The first pulse shows the first passage of the wave under the fibre and contains no information regarding the defects. The other feature which is expected is the backwall reflection and this potentially contains useful information as the wave in question would pass through the defect twice. If any direct wave scattering from a defect is present it will occur between these features. In order to compensate for the effect of uncertain coupling between the source and plate, all the patterns — scanned at 10mm increments across the plate — were normalised by dividing by the initial pulse height. The means of the signals were removed in all cases as were any spikes in the data from the instrumentation. The final stage of preprocessing was to decimate by a factor of 5, giving a 50-point feature vector for training the novelty detector.

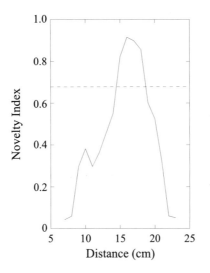

Fig. 4.13   Novelty index for test plate.

The second stage of producing the novelty index is training the AA network. The two leftmost and two rightmost data vectors were taken as normal condition. The training set was obtained by making 250 copies of each of the patterns corresponding to normal condition and corrupting each copy with different Gaussian noise vectors. (A typical (clean) pattern is shown in Fig. 4.12). The object of this exercise is to produce a novelty detector which never fires purely because a measured pattern is noisy. In the absence of any prescription for the noise, the Gaussian process (with unit-proportional covariance matrix) was chosen; a minimal requirement for any pattern recognition system is that it should be transparent to normally distributed noise. In geometrical terms, the assumption here is that the normal condition set in pattern space is spherical.

After the neural network was trained, it was presented with the patterns measured from the plate sequentially from left to right. The resulting novelty index as a function of position is given in Fig. 4.13. The fault is clearly identified. The dotted line in the graph is the novelty threshold computed from the training data and the index rises above it at the location of the fault. It was assumed that the index statistics were Gaussian and the threshold represents mean plus four standard deviations.

As discussed above, the novelty detector is a level one damage diagnostic. However, in some cases, it is possible to recover a little more information. Because the novelty index is a function of distance along the plate, the range of super-threshold values may give an indication of the width of the defect. Figure 4.13 shows that there are four values above threshold, suggesting a width of 40mm. In fact, the width is 15mm. The overestimate is due to the fact that the wavefront spreads out as it travels across the plate.

4.10 Sensor Optimisation, Validation and Failure-Safety

4.10.1 *Optimal Sensor Distributions*

The second case study illustrates a number of different applications of computation. As in the first case, the area is damage detection. However, the discussion will focus not on the diagnostic, but on the optimisation of the sensor distribution feeding information to the diagnostic. The same methods serve to establish sensor networks which are failure-safe, i.e. still produce a diagnosis in the event of sensor failure. The system under investigation has been discussed in a number of previous works [33; 34], it is a simple finite element (FE) model of a metal plate which loosely represents an aircraft skin panel. The dimensions of the plate were $300 \times 300 \times 3$mm and the material constants of aluminium were used: density $\rho = 2700$kgm^{-3}, Young's modulus $E = 70 \times 10^9$Nm^{-2} and Poisson's ratio $\nu = 0.3$.

The boundary conditions assumed two clamped (CC) and two simply-supported (S) edges as shown in Fig. 4.14. The loading was combined bending and in-plane loading. The FE modelling of the plate was carried out using the ABAQUS software package and employed 900 eight-noded quadrilateral shell elements on a regular 30×30 mesh.

For the purposes of fault location, the plate was divided into thirty six regions (Fig. 4.15 shows the numbering of the regions). Damage was simulated simply by reducing the Young's modulus of a number of elements in the required region to a negligible value. The number of elements damaged was taken as nine to represent substantial damage as in Fig. 4.15 (this is not a limitation, recall that the object of the study was to optimise the sensor distribution, not the diagnostic).

The most effective data for diagnosis were established in a previous study ([33]), to be shear strains, so these were extracted from the FE anal-

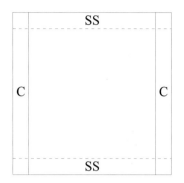

Fig. 4.14   Boundary conditions for the analysed plate.

ysis. Clearly the amount of data available for training is unrealistically large; the candidate locations for sensors were therefore assumed to be on the 5 × 5 element sub-mesh shown in Fig. 4.16.

The structure and training of the diagnostic neural network is described in detail in [33]. It has 36 outputs, one for each possible damage location. When a measurement vector is presented, the location corresponding to the highest network output is indicated. (The *1 of M* strategy was used for training, and this allowed a probabilistic interpretation of the outputs, see [35].) This means that the network was trained to produce a value of 1.0 only at the output corresponding to a given class, all other outputs being zero. Such networks actually will produce the *probability* of class membership at each output when trained.

In order to assess the effectiveness of a given sensor distribution, a diagnostic network was trained and tested and the number of misclassifications over the training set were totaled and used to estimate a probability of misclassification.

A genetic algorithm (GA) was used to determine the best sensor set or input set for damage detection, the analysis was restricted to 3-sensor distributions in order to illustrate the method. Although this meant that the effectiveness of the distributions was not good, it did allow the optimised distributions to be validated against the results of exhaustive search.

For a more detailed analysis of genetic algorithms (GAs), the reader is referred to [36]. Very briefly, GAs are search procedures based on the mechanism of natural selection. A *population* of solutions are iterated, with the fittest solutions propagating their genetic material into the next

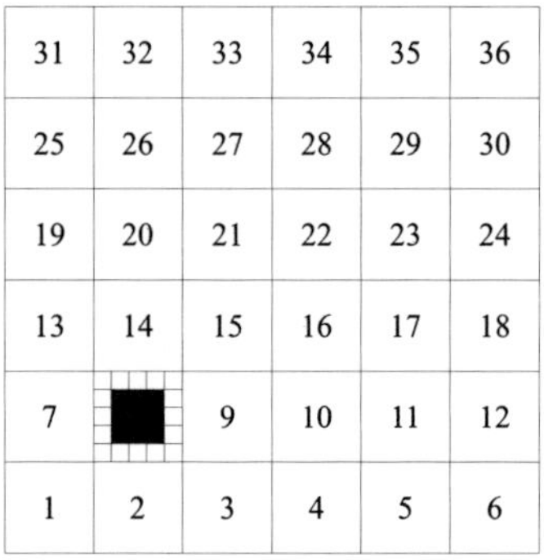

31	32	33	34	35	36
25	26	27	28	29	30
19	20	21	22	23	24
13	14	15	16	17	18
7	■	9	10	11	12
1	2	3	4	5	6

Fig. 4.15   Numbering for fault locations. The dark area indicates nine finite elements with reduced stiffness at the eighth fault location.

generation by combination with other solutions. In the simplest form of GA, each possible solution is coded into a binary bit-string which constitutes the gene. Mating is implemented by exchanging corresponding sections of pairs of genes. Mutation is also simulated by the occasional random switching of a bit.

It was shown in [34], that the binary representation is unsuitable for the sensor optimisation problem. As a result a modified GA is used, where the gene is a vector of integers, each specifying the position of a sensor; i.e. the gene (2,14,23) represents a 3-sensor distribution, with sensors at locations 2, 14 and 23 on the candidate mesh. The operations of reproduction, crossover and mutation for such a GA are straightforward modifications of those for a binary GA [37].

The initial population for the GA was generated randomly as standard. The genes — in this case sensor distributions — are propagated according to their fitness. In this case, the fitness was the inverse of the probability of misclassification of the diagnostic network.

The parameters used for the GA runs were as follows: population size of 50, number of generations 50, probability of crossover 0.8, probability of mutation 0.05. A single member elite was used and five new blood were added at each iteration. Linear fitness scaling was used. For explanations of all these terms, refer to [37] or [36].

The GA produced a best distribution (4,16,25) with a fitness of 3.155, corresponding to a probability of misclassification of 0.317 for the neural network diagnostic. The distribution is shown in Fig. 4.17.

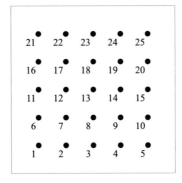

Fig. 4.16 Candidate measurement points.

In order to evaluate this distribution, it was compared with the results of an exhaustive search over all 3-sensor distributions. In fact, the exhaustive search gave a different solution, the distribution (8,16,25) was obtained with a probability of error of 0.321. This distribution (Fig. 4.18), is very close to the GA solution. The fact that the GA appears to have found a better result than any from the exhaustive search is not a problem. Each diagnostic network starts from different random initial conditions, so the probability of error is in itself subject to statistical fluctuations. Because of this, the best GA result was not the optimum according to exhaustive search. In any case, the efficiency of the GA is confirmed by this experiment.

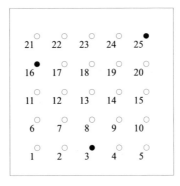

Fig. 4.17 Best 3-sensor distribution from the GA.

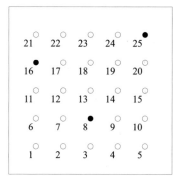

Fig. 4.18   Best 3-sensor distribution from the exhaustive search.

4.10.2   *Failure-Safe Distributions*

The same optimisation approach suffices to find the fail-safe distributions. However, the *fail-safe* fitness is obtained by a slightly more lengthy procedure. Given an N-sensor *mother* distribution, all possible $N-1$-sensor *child* distributions are generated and assessed by fitting a neural network. The fail-safe fitness of the mother is obtained by taking the *worst* probability of error of a child distribution and inverting it.

The GA was used for the 3-sensor distributions using the fail-safe fitness measure. The result was the distribution in Fig. 4.19 with a fitness of 1.704. This means that the *worst* child 2-distribution had a probability of error of 0.587. This distribution is a little surprising as the sensors are clustered

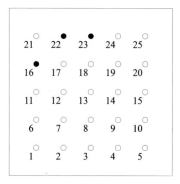

Fig. 4.19   Best fail-safe 3-sensor distribution from the exhaustive search.

quite close together. However, exhaustive search found the optimal fail-safe fitness to be 1.75, so the GA result is excellent.

In fact, the worst child of the distribution (16,22,23) is still one of the best 2-sensor distributions as exhaustive search on 2-sensor networks showed. However, the mother distribution is actually quite poor. The reason is that the GA optimised on the basis of failure-safety and made no allowance for the performance of the mother. In practice, the objective function would have to be a combination of the two fitnesses.

Sensor Validation

This final section is essentially speculation, but it seems evident that the next development in the design and implementation of sensor networks should be the inclusion of some mechanism for self-validation. Self-validation can be interpreted in two ways: in the first situation, the individual sensor contains hardware and/or software which allows it to return a *validity index* along with each measurement; this is the approach taken in the SEVA program [11; 12; 13].

The second approach assumes basic sensor technology and the validation is essentially performed within the network. Four basic approaches present themselves.

Active 1 A central actuator sends a reference signal to all sensors in the network, this can then be compared to the expected response using a novelty index.

Active 2 As **Active 1** except that in some cases, i.e. piezoelectrics, the sensors themselves can actuate and interrogate the other sensors. This allows cross-validation.

Passive 1 The response probability distribution (PDF) of the sensor is computed (the relevant technology is available [38]), subsequent measurements are inspected to see if they are outliers and therefore novel.

Passive 2 Each subset of sensors can be used to compute a conditional PDF for sensors not in the set and the approach then follows **Passive 1**.

The existence of self-validating networks is the final stage in establishing a holistic approach to damage detection. Computers are used at all stages in the process, from modelling the fault data for training purposes, to establishing neural network/pattern recognition diagnostic, to optimising sensor distributions, to validating distributions.

4.11   Conclusions

The case studies above show that, in principle, all of the desirable properties of smart structures enumerated in the introduction can be implemented in practice through the use of appropriate data fusion strategies, signal processing and computation. The main omission in the discussion is the role of active control in the establishing of smart systems. This was deliberate as any developments in active control for smart systems will arguably arise through mainstream research in that field, and in any case results can be found elsewhere in this book. Developments in optoelectronics and massively parallel computer architectures can only increase the possibilities in the future.

Acknowledgements

The authors would particularly like to thank Dr. Gareth Pierce, Dr. Wayne Philp and Prof. Brian Culshaw of the Optoelectronics Group, Department of Electrical Engineering, University of Stratclyde, for providing the Lamb wave data discussed in the first case study. They would also like to thank Susannah Side and Robin Wardle of the Dynamics Research Group for their invaluable input into the second case study.

Appendix A

The Multi-Layer Perceptron

The neural network paradigm used for this study was the multi-layer perceptron (MLP). For the sake of completeness, a brief description of the MLP is given here; for a more detailed discussion, the reader is referred to [39] or [32].

The MLP is simply a collection of connected processing elements called nodes or neurons arranged together in layers. A set of signal values pass into the *input layer* nodes, progress forward through the network *hidden layers* and the result finally emerges through the *output layer*. Each node i is connected to each node j in the preceding layer through a connection of weight w_{ij} and to nodes in the following layers. Signals pass through the node as follows: a weighted sum is performed at i of all the signals x_j from the preceding layer, giving the excitation z_i of the node; this is then passed through a nonlinear *activation function* f to emerge as the output of the node x_i to the next layer, i.e.

$$x_i = f(z_i) = f\left(\sum_j w_{ij} x_j\right). \qquad (A.1)$$

Various choices for the function f are possible; the hyperbolic tangent function $f(x) = \tanh(x)$ was used here. One node of the network, the *bias* node is special in that it is connected to all other nodes in the hidden and output layers, the output of the bias node is held constant throughout in order to allow constant offsets in the excitations z_i of each node.

The first stage of using a network to model an input-output system is to establish the appropriate values for the connection weights w_{ij}. This is the *training* or *learning* phase. The type of training adopted here is a form of *supervised* learning and makes use of a set of network inputs for which the desired network outputs are known. At each training step, a set of inputs are passed forward through the network yielding trial outputs which are then compared to the desired outputs. If the comparison error is considered small enough, the weights are not adjusted. If, however, a significant error is obtained, the error is passed *backwards* through the net and a *training algorithm* uses the error to adjust the connection weights. The algorithm used in this work is the *backpropagation* algorithm which can be summarised briefly as follows. For each presentation of a training set, a measure J of the network error is evaluated where

$$J(t) = \frac{1}{2} \sum_{j=1}^{n^{(l)}} (y_i(t) - \hat{y}_i(t))^2 \qquad (A.2)$$

and $n^{(l)}$ is the number of output layer nodes. J is implicitly a function of the network parameters $J = J(\theta_1, \ldots, \theta_n)$ where the θ_i are the connection weights ordered in some way. The integer t labels the presentation order of the training sets. After a presentation of a training set, the standard steepest descent algorithm requires an adjustment of the parameters:

$$\triangle \theta_i = -\eta \frac{\partial J}{\partial \theta_i} = -\eta \nabla_i J, \qquad (A.3)$$

where ∇_i is the gradient operator in the parameter space. The parameter η determines how large a step is made in the direction of steepest descent and therefore how quickly the optimum parameters are obtained. For this reason η is called the learning coefficient. Detailed analysis [39] gives the update rule after the presentation of a training set:

$$w_{ij}^{(m)}(t) = w_{ij}^{(m)}(t-1) + \eta \delta_i^{(m)}(t) x_j^{(m-1)}(t), \qquad (A.4)$$

where $\delta_i^{(m)}$ is the error in the output of the ith node in layer m. This error is not known *a priori* but must be constructed from the known errors $\delta_i^{(l)} = y_i - \hat{y}_i$ at the output layer. This is the source of the name backpropagation, the weights must be adjusted layer by layer moving backwards from the output layer.

There is little guidance in the literature as to what the learning coefficient η should be; if it is too small, convergence to the correct parameters may be very slow. However, if η is made large, learning is much more rapid but the weights may diverge or oscillate. One way around this problem is to introduce a *momentum* term into the update rule so that previous updates persist for a while, i.e.

$$\triangle w_{ij}^{(m)}(t) = \eta \delta_i^{(m)}(t) x_j^{(m-1)}(t) + \alpha \triangle w_{ij}^{(m)}(t-1), \qquad \text{(A.5)}$$

where α is termed the momentum coefficient. The effect of this additional term is to damp out high frequency variations in the backpropagated error signal.

Once the comparison error is reduced to an acceptable level over the whole training set, the training phase ends and the network is established. The networks used for this study were designed and trained using the package MLP [40].

Bibliography

[1] Klein, L. A. (1999) *Sensor and Data Fusion: Concepts and Application*, SPIE Press.

[2] Esteban, J. and Starr, A. G. (1999) "Building a Data Fusion Model" *Proc. Int. Conf. on Data Fusion — EuroFusion 99, Stratford-upon-Avon, UK*, pp. 187–196.

[3] Bedworth, M. and O'Brien, J. (1999) "The Omnibus Model: A New Model of Data Fusion" (preprint).

[4] Sandell, N. R., Lauer, L. C. and Kramer, L. C. (1980) "Research Issues in Surveillance for C^3" *Proc. 19th IEEE Conf. on Decision and Control*, pp. 201–213.

[5] Athans, M. (1980) "System Theoretic Challenges and Research Opportunities in Military C^3 Systems" *Proc. 19th IEEE Conf. on Decision and Control*, pp. 12–16.

[6] White, F. E. Jr. (1990) *Technical Proceedings of the Joint Service Data Fusion Symposium*, I-DFS-90, Joint Directors of Laboratories Data Fusion Subpanel report: SIGINT session, pp. 469-484.

[7] Data Fusion Development Strategy Panel (1991) *Functional Description of the Data Fusion Process*, Office of Naval Technology.

[8] Waltz, E. and Llinas, J. (1990) *Multisensor Data Fusion*, Artech House.

[9] Rytter, A. (1993) *Vibration Based Inspection of Civil Engineering Structures*, PhD thesis, Department of Building Technology and Structural Engineering, University of Aalborg, Denmark.

[10] Gatepaille, S., Brunessaux, S. and Abdulrab, H. (1999) "Data Fusion Multi-Agent Framework" *Proc. Int. Conf. on Data Fusion — EuroFusion 99, Stratford-upon-Avon, UK*, pp. 97–102.

[11] Henry, M. P. and Clarke, D. W. (1993) "The Self-Validating Sensor: Rationale, Definitions and Examples", *Control Eng. Practice*, **1**, 585–610.

[12] Henry, M. P. (1994) "Validating Data from Smart Sensors", *Control Eng.*, **41**, 63–66.

[13] Henry, M. P. (1995) "A Self-Validating Coriolis Meter", *Control Eng.*, **42**, 81–86.

[14] Side, S., Staszewski, W. J., Wardle, R. and Worden, K. (1997) "Fail-safe sensor distributions for damage detection" *Proc. Int. Workshop on Damage Assessment Using Advanced Signal Processing Procedures — DAMAS 97, Sheffield, UK*, pp. 135–146.

[15] Boyd, J. (1987) *A Discourse on Winning and Losing*, Maxwell AFB lecture.

[16] Bossé, É. and Bertrand, S. (1999) "R&D Perspectives on Data Fusion and Decision Support Technologies for Naval Operations" *Proc. Int. Conf. on Data Fusion — EuroFusion 99, Stratford-upon-Avon, UK*, pp. 85-95.

[17] Bedworth, M. (1994) "Probability Moderation for Multilevel Information Processing", DRA technical report DRA/CIS(SE1)/651/8/ M94.AS03BP032/1.

[18] Hall, D. L. and Linn, R. J. (1993) "Survey of Commercial Software for Multi-Sensor Data Fusion" *Proc. SPIE Conf. on Sensor Fusion and Aerospace Applications*, **1956**, pp. 98–109.

[19] Viktorov, I. A. (1967) *Rayleigh and Lamb Waves: Physical Theory and Applications*, Plenum Press.

[20] Alleyne, D. N. and Cawley, P. (1992) "The Interaction of Lamb Waves with Defects", *IEEE Trans. on Ultrasonics, Ferroelectrics and Frequency Control*, **39**, 381–397.

[21] Alleyne, D. N. and Cawley, P. (1992) "Optimisation of Lamb Wave Inspection Techniques", *NDT&E Int.*, **25**, 11–22.

[22] Cawley, P. and Alleyne, D. N. (1996) "The Use of Lamb Waves for the Long Range Inspection of Large Structures", *Ultrasonics*, **34**, 287–290.

[23] Guo, N. and Cawley, P. (1994) "Lamb Wave Reflection for the Quick Non-Destructive Evaluation of Large Composite Laminates", *Mat. Eval.*, **52**, 404–411.

[24] Jansen, D. P., Hutchins, D. A. and Mottram, J. T. (1994) "Lamb Wave Tomography of Advanced Composite Laminates Containing Damage", *Ultrasonics*, **32**, 83–89.

[25] Pierce, S. G., Philp, W. R., Culshaw, B., Gachagan, A., McNab, A., Hayward, G. and Lecuyer, F. (1996) "Surface-Bonded Optical Fibres for the Inspection of CFRP Plates Using Ultrasonic Lamb Waves", *Smart Mat. & Struct.*, **5**, 776-787.

[26] Staszewski, W. J., Pierce, S. G., Worden, K., Philp, W. R., Tomlinson, G. R. and Culshaw, B. (1997) "Wavelet Signal Processing for Enhanced Lamb Wave Defect Detection in Composite Plates Using Optical Fibre Detection", *Optic. Eng.* (to appear).

[27] Bishop, C. M. (1995) *Neural Networks for Pattern Recognition*, Oxford University Press.

[28] Tarassenko, L., Hayton, P., Cerneaz, N. and Brady, M. "Novelty Detection for the Identification of Masses in Mammograms" (preprint).

[29] Surace, C., Worden, K. and Tomlinson G. R. (1997) "A Novelty Detection

Approach to Diagnose Damage in a Cracked Beam", *Proc. 15th Int. Conf. on Modal Analysis, Orlando, Florida*, pp. 947–953.

[30] Worden, K. (1997) "Structural Fault Detection Using a Novelty Measure", *J. Sound & Vib.*, **201**, 85–101.

[31] Pomerleau, D. (1993) "Input Reconstruction Reliability Estimation", *Advances in Neural Information Processing Systems 5* (ed. S. J. Hanson, J. D. Cowan and C. L. Giles), Morgan Kaufman.

[32] Rumelhart, D. E. and McClelland, J. L. (1988) *Parallel Distributed Processing: Explorations in the Microstructure of Cognition*, MIT press.

[33] Staszewski, W. J., Worden, K. and Tomlinson G. R. (1996) "Optimal Sensor Placement for Neural Network Fault Diagnosis", *Proc. Adaptive Computing in Engineering Design and Control 96*, pp. 92–99.

[34] Worden, K. and Burrows A. P. (1997) "Optimal Sensor Location for Fault Detection", *Eng. Struct.* (to appear).

[35] Richard, M. D. and Lippmann, R. P. (1991) "Neural Network Classifiers Estimate Bayesian *a Posteriori* Probabilities", *Neural Computation*, **3**, 461–483.

[36] Goldberg, D. E. (1989) *Genetic Algorithms in Search, Optimization, and Machine Learning*, Addison–Wesley.

[37] Staszewski, W. J. (1995) "Optimal Sensor Distributions for Fault Diagnostics Part II: Neural Network Analysis", report DCRG-BAe-2/95, Department of Mechanical Engineering, University of Sheffield, Sheffield, UK.

[38] Silverman, B. W. (1986) *Density Estimation for Statistics and Data Analysis*, Chapman and Hall.

[39] Billings, S. A., Jamaluddin, H. B. and Chen, S. (1991) "A Comparison of the Backpropagation and Recursive Prediction Error Algorithms for Training Neural Networks", *Mech. Sys. & Signal Process.*, **5**, 233–255.

[40] Worden, K. (1996) *MLP 3.4 — A User's Manual* (obtainable via k.worden@sheffield.ac.uk).

Chapter 5

Shape Memory Alloys — A Smart Technology?

Neil B. Morgan and Clifford M. Friend

Department of Materials and Medical Sciences,
Cranfield University,
Shrivenham, Swindon SN6 8LA, UK.

5.1 Introduction

Shape memory alloy (SMA) is the generic name given to alloys which exhibit the unusual property of a strain-memory which can occur either at constant temperature, where large yet recoverable strains are possible (superelasticity), or on changes in temperature, where apparently permanent strains can be fully recovered (thermal shape memory).

These memory effects have their origin in a particular type of phase transformation (change in internal crystal structure) which produces a microstructural constituent known as martensite. A martensitic transformation is displacive, that is it occurs through a shearing of the crystal structure from the so-called parent-phase to that of the martensite. This is illustrated schematically in the two-dimensional analogue shown in Fig. 5.1. Such transitions are diffusionless, resulting in no change in chemical composition, and form the basis of heat treatments in many metallic materials, including the familiar transformation of austenite to martensite in ferrous (iron-based) alloys. Thermally induced martensitic transformations occur over a falling temperature interval (Fig. 5.2). The transformation proceeds at the M_s (martensite start temperature) where the martensite phase first appears, and progresses athermally (occurring over a falling temperature interval)

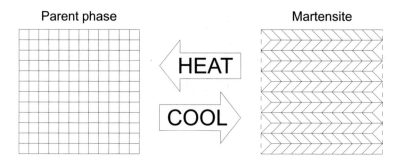

Parent phase   Martensite

Fig. 5.1   Two-dimensional, two-variant analogue of a thermoelastic martensitic transformation.

until at the M_f temperature transformation is complete. On heating, at the A_s temperature the last martensite formed during forward transformation begins to revert to parent-phase, and there is then continuous reversion until at the A_f temperature the high temperature parent-phase is fully restored. The overall hysteresis between forward and reverse transformation pathways in SM alloys is small, usually between 10 and 50°C; a behaviour quite different to that exhibited by non-thermoelastic martensitic transformations, typified by steels. Thermoelastic martensitic transformations, which form the basis of SMA behaviour, can be repeated indefinitely as long as high temperature excursions are avoided.

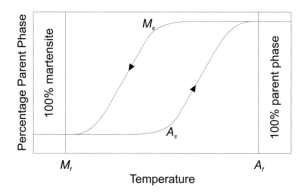

Fig. 5.2   Hysteresis curve for a thermoelastic martensitic transformation.

5.2 Structural Origins of Shape Memory

A feature of all martensitic transformations is that there are a number of equivalent shear directions through which the martensite can form within a region of parent-phase. This results in the formation of martensite *variants* within the microstructure of a transformed alloy. This is illustrated again schematically in Fig. 5.1 using a two-dimensional, two-variant model. This figure shows two crystallographically equivalent martensite variants created by different atomic shears from the parent phase. In this analogue the two opposite shears maintain the macroscopic shape of the crystal block (represented by the dotted line). Such a microstructure, where the shear of one variant is accommodated or "cancelled" by that of the other, is known as a self-accommodated structure. This process forms the basis of the shape memory effect in SMAs, although three-dimensional self-accommodation requires a larger number of variants (typically up to 48 in many alloys).

It is apparent from the two-dimensional analogue in Fig. 5.1 that the parent phase of a SMA has higher symmetry than the martensite. This means that although there are many transformation routes through which the martensite can form from the parent phase, there are only a few possible routes for the reverse transformation; often limited by other factors to a single reversion pathway. In other words transforming to martensite and reverting to parent phase results in a complete restoration of microstructure (Fig. 5.1). In SMAs the interfaces between martensite variants are glissile (mobile) and their positions can be influenced by external variables; perhaps most importantly by applied stress. This is illustrated in Fig. 5.3 where the positions of the martensite interfaces change under the influence of stress/strain, creating a balance of variants whose shears best accommodate the direction of applied strain. That is the interfaces between variants move to "grow" the most favourably oriented variants and shrink the least. It is the ability to re-orientate martensite variants by the application of stress and to revert these to parent-phase which forms the basis of at least one shape memory phenomenon — the so-called one-way shape memory effect.

5.3 One-Way Shape Memory

Figure 5.4 schematically illustrates the macroscopic response of a one-way SMA. If such an alloy is deformed whilst in the self-accommodated marten-

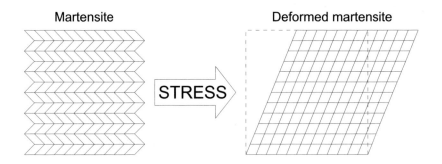

Fig. 5.3   Variant reorientation in thermoelastic martensites.

sitic state and subsequently unloaded, then an apparently permanent strain will remain. This is a result of the martensite microstructure being re-oriented as shown in Fig. 5.3; the reorientation remaining on the removal of the external stress. If the alloy is then reheated to a temperature above the martensitic transformation temperature range then this apparently permanent strain will fully recover, returning the original macroscopic shape.

This is the so-called one-way memory effect. As long as the total strain does not induce permanent plastic flow, deformation may be of any type, e.g. tension, compression, bending or more complex combinations.

The internal structural changes that take place during the one-way memory effect can be visualised using the two-dimensional analogue in Fig. 5.5. Deformation takes place in the self-accommodated martensite

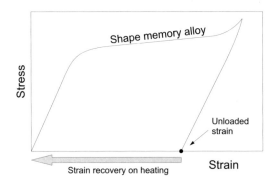

Fig. 5.4   Stress–strain behaviour during the one-way memory effect.

condition. During loading this structure becomes deformed through variant rearrangement, resulting in a net macroscopic shape change. When the alloy is unloaded this deformed structure remains, resulting in an apparent permanent strain. However, if the alloy is now reheated to a temperature above the martensitic transformation range the original parent phase microstructure and macroscopic geometry is restored. This is possible because no matter what the post-deformation distribution of martensite variants, there is only one reversion pathway to parent-phase for each variant. When the alloy is subsequently cooled to below the transformation range a self-accommodated martensite microstructure is formed and the original shape before deformation is retained. Thus a one-way shape memory is achieved. The maximum strain recovered through this process depends on the shape memory system, however, it is typically in the range 1–7% for polycrystalline alloys.

5.4 Two-Way Memory Effect

During the one-way memory effect only one shape is 'remembered' by the alloy; the so-called hot (parent phase) shape. However, SMAs can be processed to remember both hot and cold shapes, thus exhibiting a two-way memory where the component can be cycled between two different shapes without the need for an external stress.

Two-way shape changes rely entirely on microstructural changes during martensitic transformation which occur under the influence of internal stress [1]. Self-accommodation of the martensite microstructure is lost in the two-way effect due to the presence of these internal stresses, and predominant variants form during transformation (i.e. there is an excess of certain variants within the martensite microstructure compared to self-accommodated structures). This results in a shape change towards the cold-shape on cooling and towards a second hot (parent phase) shape on heating through the reverse transformation. This cycle is also illustrated in Fig. 5.5. Notice how the self-accommodated structure is missing in such alloys and that the martensite shape is achieved directly by cooling below the M_s temperature under the influence of internal stress. Internal stress may be introduced in a number of ways, usually referred to generically as "training". These sites of stress must be stable on thermal cycling through the transformation and usually result from the introduction of irreversible de-

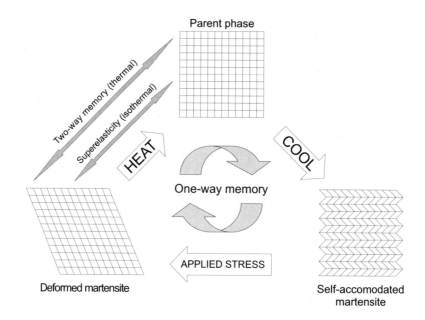

Fig. 5.5   Microstructural changes during thermal memory and superelastic phenomena.

fects. These are created by prior deformation [2; 3] or through the presence of particles and precipitates [4] created during special thermo-mechanical treatments [5]. Two of the most common training methods create two-way memory through the introduction of dislocation arrays and are achieved by:

- Cyclic deformation at a temperature below M_f followed by constrained heating in the cold-shape to a temperature above A_f.
- Cyclic deformation between the hot- and cold-shapes at a temperature above A_f [6].

5.5   Pseudoelasticity or the Superelastic Effect

When a SMA is deformed isothermally at a temperature above A_f, martensitic transformation can also be induced mechanically. The martensite formed in this way is known as stress-induced martensite (SIM). This is only stable under the application of stress, and on unloading, the reduction in stress and surrounding elastic forces generated during transformation cause the martensite to shrink back to the original parent-phase. Figure 5.6 shows the mechanical behaviour of such a superelastic material, and compares this to that of a conventional metallic alloy. Such superelastic materials can fully recover deformations up to 6–7% strain, depending on alloy type. It can be seen from Fig. 5.6 that superelastic deformation is also hysteretic, the upper plateau occurring during stress-induced martensitic transformation and the lower during reversion on unloading. It is both the large recoverable strain and constant recovery stress plateau that can be utilised in superelastic applications.

Figure 5.5 can also be used to explain the microstructural origin of these effects, with stress-induced transformation resulting in a predominant variant microstructure ("deformed martensite structure"), creating a macroscopic strain, which shrinks on reversion of the martensitic phase. The deformation hysteresis is also clearly associated with the inherent hysteresis of the underlying martensitic transformation (Fig. 5.2).

5.6 A Brief History of Memory Alloys and their Application

Arguably the first observations of shape memory behaviour were carried out by Ölander [7] in 1932 in his study of a "rubber like effect" in the Au–Cd system and by Greninger and Mooradian [8], 1938, in their study of Cu–Zn alloys. However, it was many years later that Chang and Read [9] first reported the term "shape recovery" whilst working on Au–Cd alloys. It was not until 1963, in a study on NiTi alloys, that Buehler *et al.* [10] first introduced the phrase "shape memory effect" as a material property. Indeed it was the discovery of the effect in these NiTi alloys that "kick-started" interest in shape memory applications. During the 1960s NiTi alloys and their early applications began to move the effect away from fundamental phenomena to useful engineering properties and fuelled international research. Duerig [11] divides the methods of harnessing thermal memory effects into three categories:

Free Recovery An alloy is apparently permanently strained and on the application of heat recovers its original shape; maintaining this during subsequent cooling. The function of the alloy element is therefore to cause motion or strain.

Constrained Recovery The alloy is prevented from full shape recovery thus generating stress on the constraining element.

Actuation Recovery The alloy is able to recover its shape but operates against applied stress, resulting in work production.

Duerig [11] also considers superelastic applications in the following way:

Superelastic Recovery The only isothermal application of the memory effect, superelastic recovery (also known as pseudoelasticity) involves the storage of potential energy through comparatively large but recoverable strains.

In some cases separation of the thermal and isothermal applications is not possible since many superelastic applications also involve free and constrained recovery (where the superelastic element either recovers its shape freely or is constrained by another mechanical element of a device). These terms therefore need not be confined to thermal effects. However, this broad type of categorisation is useful for conceptualising how shape change phenomena may be applied within engineering systems.

Although there are examples of NiTi SMA applications in all of the four categories described by Duerig [11], the greatest number have emerged in the area of superelasticity. Many excellent and unique devices have been constructed out of NiTi for the medical industry and this market is still growing at a considerable pace. It is perhaps, the reported biocompatability [12; 13; 14] allied to the less complicated design procedures for superelastic applications that has resulted in the high number of devices utilising this effect.

Guide wires for non-invasive surgery [15], orthodontic arch wires [16; 17] highly flexible surgical tools [18] and stents for the minimally invasive treatment of arterial and oesophageal strictures [19; 20] have all been successfully produced out of NiTi superelastic wires. In addition to the medical market, superelastic components are also successfully employed in consumer products such as spectacle frames [21] under-wired bras [22] and mobile phone antennae [23].

The three categories of thermal shape recovery have met with much less commercial success and are limited to just a few niche areas. Of these, perhaps the most successful is also one of the first. This is the coupling originally designed by the Raychem Corporation to employ constrained recovery for the joining of pipes in the Grumman F-14 aircraft [24]. In this application a ring of NiTiFe alloy is expanded in diameter at very low

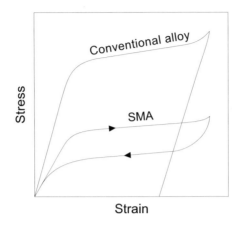

Fig. 5.6   Comparison of the stress–strain curves of conventional and superelastic alloys.

temperatures and fitted over the coupling area of the pipes. As the ring returns to comparatively high ambient temperatures it attempts to contract to its original diameter and exerts high forces on the pipes resulting in very strong, reliable couplings. As well as pipe couplings SMAs have also been used in various electrical connectors and fastener applications, all utilising constrained recovery effects [25; 26].

Unconstrained recovery applications are few. Devices exploiting shape change only include fire protection devices and thermal cut out switches [27]. In these applications the alloy element acts as both thermal sensor and cut-out actuator and tend to be concerned with flow cut off valves. For instance the Proteus gas valve was specifically designed to cut off gas flow in the case of fire [28]. In this product a CuZnAl SMA spring expands at a particular temperature pushing a steel ball through a retaining ring. The valve can be reset manually at temperatures below the SMAs martensitic transformation and does not have to undergo repeated actuation or operate against constant loads.

The area of thermal shape change that perhaps has the greatest potential, for smart applications, is that of actuation [29; 30]. Many patents exist based on the principle of using a SMA element as a thermal actuator which converts electrical or thermal energy into mechanical work. However, because of the complicated design criteria for matching the desired motion, cycle life and actuation temperatures to an "off the shelf" material [31;

32] the number of successful repeatable actuation devices is limited. The devices that have met with commercial success are usually those that have been designed allied to stringent actuator research and development programmes. A good example of this is the air conditioner actuator developed by Matsushita Electrical Industrial Company Ltd. on the basis of the fundamental research carried out by Todoroki [33]. It is obvious, however, that the cost of this type of research is high and many SMA actuator designs do not get any further than the conceptual stage.

5.7   Why Not Use Bimetals?

This question is often asked of shape memory researchers and manufacturers when considering possible thermal actuator applications and deserves to be seriously answered. The answers are clear. The displacements of bimetallic strips tend to be much smaller than those of shape memory alloys and vary linearly with temperature, rather than the switch-like behaviour associated with SMAs over their relatively narrow transformation temperature range. In addition SMAs may be configured into many different shapes, e.g. a spring or tubular cross section, and exhibit a tailorable direction of deformation with temperature. Finally, and particularly important for thermal actuator applications, SMAs can exert recovery forces up to 100 times greater than bimetallic strips.

5.8   Types of Shape Memory Alloy

Despite a growing list of alloys that display the memory effect only copper-based and NiTi-based alloys have been commercially exploited, and by far the most important commercial shape memory alloys are those based on the NiTi system. It is their comparatively large shape memory properties and excellent corrosion resistance [34] that really sets these alloys apart in terms of commercial application.

In all SMAs careful processing and alloying permits close control of properties such as actuation temperature (phase transformation temperatures), strength and work outputs.

Copper-based shape memory alloys exhibit higher actuation temperatures (approximately in the range -200 to $+200°C$) than NiTi alloys and are sometimes the only choice for high temperature applications, (i.e. $> 100°C$).

The practical recoverable strain in polycrystalline copper-based SMAs also ranges from approximately 3% in Cu–Al–Ni alloys to 4% in Cu–Zn–Al alloys. Unfortunately, these copper alloys tend to suffer from low strength and poor corrosion resistance.

NiTi alloys exhibit by far the greatest recoverable strains of commercially available polycrystalline shape memory alloys but generally have a lower range of actuation temperature (approximately in the range −200 to +100°C). Fully recoverable strains of 7% are easily achievable with these alloys and their comparatively high strength and excellent corrosion resistance has resulted in many unique shape memory applications. The poor electrical conductivity of NiTi alloys also allows them to be used in solid-state actuator applications where the alloy is heated by electrical current. Because of this, recent research into smart structures incorporating solid-state actuators has resulted in concurrent development activities on NiTi alloys. The next section will consider NiTi alloys in greater detail.

5.9 Nickel Titanium Shape Memory Alloys

5.9.1 *Background*

Because of its discovery by Buehler *et al.* [35; 36] at the Naval Ordnance Laboratory in California, USA, NiTi alloys are often referred to as *Nitinol* (*NiTi Naval Ordnance Laboratory*). The shape memory effect is only present in binary NiTi alloys over a very narrow compositional range based around 50 atomic percent, i.e. 50% nickel, 50% titanium. Differences of just 0.1 atomic percent can easily change transformation temperatures by 20°C or more. For this reason production and processing of NiTi alloys must be very strictly controlled. Induction melting is often used to produce the final ingot, ensuring good homogeneity of the alloy, and enabling transformation temperatures to be controlled to within 5°C. Unfortunately this type of careful fabrication and often small production runs adds to the cost of the final product. Because of this NiTi alloys are often regarded as being comparatively expensive.

5.9.2 *Mechanical Behaviour*

The deformation behaviour of NiTi alloys depends heavily on ambient temperature, the phase of the alloy and its transformation temperatures. This

is illustrated in Fig. 5.7. The different stages of deformation are labelled
from 1 to 4 in each figure.

Figure 5.7(a) represents the behaviour of an alloy in the parent-phase
at a temperature above M_d (the temperature above which martensite cannot be stress-induced). The curve exhibits linear deformation (stage 1)
up to approximately 0.5% strain (elastic deformation) followed by permanent (plastic) deformation (stage 2). This type of behaviour is typical of
conventional metallic alloys.

An alloy tested at a temperature above A_f but below M_d is shown in
Fig. 5.7(b). Stage 1 shows some initial elastic loading of the parent-phase.
At a particular stress martensite is induced and further strain results in
stress-induced martensitic transformation. If the alloy is unloaded at this
stage superelastic shape recovery will take place and the curve will follow
the path represented by the broken line. However, if the alloy is strained

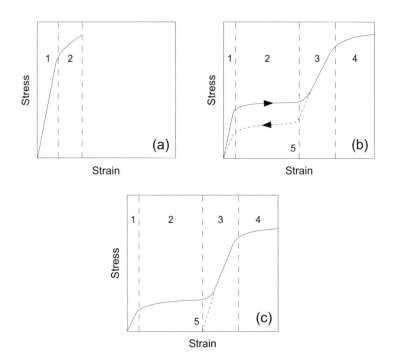

Fig. 5.7   Deformation behaviour of NiTi shape memory alloys at a temperature: (a)
above M_d; (b) above A_f but below M_d; (c) below M_f.

further, elastic deformation of the stress-induced martensite (stage 3) will occur until permanent (plastic) deformation sets in (stage 4).

Figure 5.7(c) represents an alloy tested below its M_f temperature in the fully self-accommodated martensite condition. After elastic deformation of the martensite at very low stresses (stage 1), martensite variants begin to realign at a constant stress until reorientation is complete (stage 2). Subsequent loading results in elastic deformation of the martensite variants (stage 3) and eventual permanent plastic deformation (stage 4). If the alloy is unloaded at the end of stage 2 or during stage 3, the material will elastically unload with an apparently permanent strain (5), which can be recovered by heating through the one-way shape memory effect.

All phenomena are observable in a single SMA, depending on the position of ambient temperature with respect to the alloy's transformation temperatures.

5.9.3 *Corrosion Characteristics*

One of the differentiating benefits of NiTi over other commercial shape memory alloys is its excellent corrosion resistance. The passive titanium dioxide (TiO_2) surface film results in a corrosion resistance comparable to 316L stainless steel. Formation of this film is exactly the same as for pure titanium. The film is very stable and resistant to many forms of potentially corrosive attack, however, breaks in the surface can be slow to recover.

NiTi's corrosion resistance has led to extensive studies on the bio-compatibility of these alloys and a number of medical applications. It has been established through clinical tests that the bio-compatibility is excellent, with organs showing no signs of metallic contamination. It is thought that the titanium-rich oxide surface prevents potentially harmful nickel reaching tissues. This has resulted in great activity in the medical market which has been one of the catalysts driving recent shape memory research.

5.9.4 *Ternary Additions*

The addition of a third chemical element to NiTi alloys can provide control over particular characteristics of the shape memory effect. For instance:

- Grumman F-14 pipe couplings were made from NiTiFe to provide a very low M_s temperature.

- NiTiCu alloys have become increasingly popular because of their smaller transformation hysteresis and improved cyclic stability.
- The addition of palladium to NiTi presents the interesting possibility of using NiTi-based SMAs in higher temperature applications such as automobiles because of their relatively high transformation temperatures.
- NiTiNb alloys exhibit a wide hysteresis which is particularly useful for coupling devices. This is because it is possible to deform couplings at liquid Nitrogen temperatures (−196°C) and store them in the deformed state at room temperature. Upon subsequent heating the couplings will contract and maintain the clamping pressure when cooled back down to room temperature.

5.9.5 *Summary of Mechanical and Physical Properties*

NiTi alloys offer many unique memory properties. Unfortunately, their comparatively high cost and poor workability is often a factor in determining whether an application has commercial feasibility. For reference and comparison, Tables 5.1 to 5.4 summarise the physical, mechanical and commercial properties of NiTi and copper-based shape memory alloys.

5.10 NiTi Shape Memory Alloys in Smart Applications

Soon after Buehler first discovered the shape memory effect in NiTi alloys the commercial world began to try and assimilate memory characteristics within real products. It is perhaps, because NiTi shape memory effects were first realised in the 1960s that so much initial effort went into finding applications. During this decade, and the one before it, new materials were heralded as the key to economic growth and commercial success throughout many industrial sectors. The impact of thermoplastics, semiconductors and new metallic alloys resulted in this period being referred to as the materials revolution and the discovery of a metal which changes its shape was seen as yet another material that would 'shape' the future of industry. Of course the effect had already been found in copper-based systems but had so far only been viewed as a physical curiosity that could be used to learn more about martensitic transformations.

Table 5.1   Physical properties of shape memory alloys.

	NiTi	Cu-Based Alloys
Density (gcc^{-1})	6.4–6.5	7.1–8.0
Melting point (°C)	1250	950–1050
Thermal conductivity $(Wm°C^{-1})$		
Martensite	8.6-10.0	-
Parent phase	18	79–120
Electrical resistivity $(\times 10^{-6}\Omega m)$		
Martensite	0.5–0.6	0.14
Parent phase	0.82–1.1	0.07
Co. thermal expansion $(\times 10^{-6°}C^{-1})$		
Martensite	6.6	16.0–18.0
Parent phase	10.0–11.0	-
Specific heat capacity $(Jkg°C^{-1})$	470–620	390–440
Enthalpy of transformation (Jkg^{-1})	19.0–28.0	7.0–9.0
Transformation temperature range (°C)	$-200--120$	$-200-+200$
Corrosion performance	Excellent	Poor
Bio-compatibility	Excellent	Assumed poor

In addition to the discovery of NiTi alloys at the height of a new *materials revolution*, it was also the geographical location of its discovery that resulted in such commercial interest. During the 1960s California was, and arguably still is, the innovation centre of the world. Undoubtedly, NiTi alloys rode the wave of industrial optimism in the 1960s and innovative engineers from many industrial sectors struggled to find applications for it.

However, by 1971, the time that Grumman pipe couplings finally came to market, the strategies for innovation were changing. A new model began to take shape placing more emphasis on the role of the market place. Empirical results based on real innovations began to establish a 'market pull' strategy sometimes referred to as 'need pull' [37]. It began to be realised that the most successful innovations result from either perceived or clearly defined customer needs, resulting in closely focussed research and development. However, shape memory alloy manufacturers continued with a technology 'push' strategy well into the 1980s leaving behind them a trail of failed products such as the thermally activated greenhouse latch [38].

In the early 1980s, at last a genuine market pull began to develop within the shape memory industry. Slowly it became apparent that significant numbers of papers were being presented on medical applications. Shape

Table 5.2   Mechanical properties of shape memory alloys.

	NiTi	Cu-Based Alloys
Young's modulus (GPa)		
Martensite	28–41	70
Parent phase	70–97	70–100
Yield Strength (MPa)		
Martensite	70–140	80–300
Parent phase	195–690	150–350
Ultimate tensile strength (MPa)		
Fully annealed	895	400
Work hardened	1900	1000
Elongation at failure (%)		
Fully annealed	25–50	8–15
Work hardened	5–10	8–15
Hot workability	Poor–fair	Very good
Cold workability	Poor	Good
Machinability	Poor	Very good
Poisson ratio	0.33	-
Wear resistance	Good	-

memory manufacturers and medical device companies who recognised the potential value of this market began to file strategic patents and develop products that provided unique solutions to medical problems. The market pull on shape memory applications gradually intensified during the 1980s until in 1994 the first International Conference on Shape Memory and Superelastic Technologies took place concentrating on practical applications. The influence of the medical market pull was obvious from the number of papers presented on bio-compatibility and medical applications. This conference has now become a regular event and the first European version took place in 1999.

Around the beginning of the 1990s a new market for shape memory alloys also began to emerge, that of "smart materials" and "smart structures". Whilst not being a specific, tangible market segment like that of the medical industry, recent interest in smart technologies has resulted in considerable market pull on shape memory alloy research and development.

Proactive research surrounding smart technologies has produced its own specific journals and conferences and has resulted in a resurgence of interest in using shape memory alloys in actuator applications. For this rea-

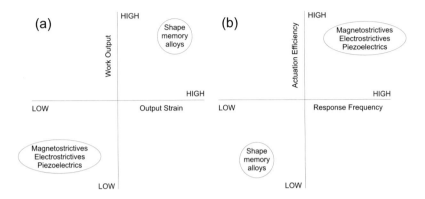

Fig. 5.8   Performance of shape memory alloys: (a) advantages; (b) disadvantages.

son it has become an important strategic segment for shape memory alloy manufacturers and a major driver of current shape memory research and development.

5.11   Shape Memory Alloys as Smart Actuators

It is clear that the main use of shape memory alloys within smart technologies will be as actuators operating through the conversion of thermal energy into motion and work output.

The use of these alloys as solid state actuators offers many benefits [39]:

- High recovery forces.
- Large recoverable output strains.
- Different actuation modes (linear, bending, torsion).
- High work output per unit volume or mass.

It is perhaps the large strains and high work output that offer most opportunities for innovative design with SMAs. The *material function* of the shape memory effect differentiates itself clearly from other actuating materials such as piezoelectrics and magnetostrictives, as shown in Fig. 5.8(a). However, when considering the benefits of SMAs the potential shortcomings should also be considered. Figure 5.8(b) shows how compared to other possible actuators, SMAs suffer from low efficiency (i.e. high loss actuation) and very poor response times or actuation frequencies. Successful applications build on SMA strengths whilst effectively designing out these weaknesses.

The use of NiTi alloys in smart structures is at an embryonic stage. Whilst many researchers are working on applying NiTi to actuator applications in smart structures, fundamental problems for their commercial use remain. These include control paradigms and thermo-mechanical fatigue effects. However, major commercial and academic players continue to research SMAs. If a new high volume application can be found then growth in the SMA "smart sector" will follow.

Much is said about the potential of the smart materials/structures market. For instance, Thompson and Gandhi [40] predict:

> "Smart materials and structures technologies will revolutionise a broad segment of the international market place for products in the defence, aerospace, automotive and commercial-products industries... The total market share of smart materials and structures is projected to exceed $65,000,000 by the year 2010."

Political, economic, social and technological forces will influence this smart materials/structures market and may ultimately dictate whether SMAs prove to be successful in creating innovation and competitive advantage. Thoughts on these key factors are highlighted below [41; 42; 43].

5.11.1 *Political Factors*

Reductions in defence budgets will result generally in less government funded R&D worldwide. This may either have positive or negative effects on SMA research. Although defence-funded R&D is likely to reduce, there is a trend towards increased funding for smart research in general. This has currently renewed interest in SMA technology within the defence market. Government funding for space research will also continue. The trend towards remote vehicle and satellite exploration will provide opportunities for SMA actuators.

5.11.2 *Economic Forces*

Aerospace sales are falling and this in turn exerts economic pressure on aerospace manufacturers. Traditionally these companies have been innovators in many areas of smart research including SMAs. If commercial applications of SMA-based smart structures are not realised soon then funding

Table 5.3 Memory properties of shape memory alloys.

	NiTi	Cu-Based Alloys
Transformation temperature range (°C)	−200−−120	−200−+200
Hysteresis (°C)	20–50	15–20
One-way memory maximum (%)	8	4–6
Two-way memory maximum (%)	3–5	1–4
Superelastic strain maximum (%)	8	2
Work output (Jg^{-1})	1–4	1

of SMA research within these industries is likely to fall sharply over the next 5 to 10 years. The continual push for increased component lifetime and cost reduction may, however, result in a positive drive towards smart technologies which have sensory and adaptive capabilities. It may be that SMAs can find niche applications within these products.

5.11.3 *Social Forces*

Increasingly stringent product safety requirements are likely to exert considerable market pull on sensory and actuator technologies including SMAs. Advances in human control environments are also exerting considerable market pull on these technologies. The trend towards interactive computer interfaces and virtual environments is putting pressure on conventional structural materials to compete with advances in computer hardware and software. Adaptive/smart materials are seen as one way of adding functionality and improving the way we relate to environments and product interfaces. SMAs are one of the few materials already available which satisfy these requirements.

Table 5.4 Economic properties of shape memory alloys.

	NiTi	Cu-Based Alloys
Composition control	Very strict	Fair
Unit cost	High	Fair
Forming cost	High	Fair

5.11.4   *Technological Forces*

Advances in enabling technologies (neural nets, sensors, etc.) will address some of the difficulties associated with the control and implementation of smart structures. In particular, the control of partially transformed SM actuators will become better understood resulting in the possibility of using SMAs as actuators in smart structures. Within the last 5 years there have also been considerable advances in NiTi product geometries such as thin films and tubing. These 'new' product forms will lead to greater freedom when designing adaptive structures and actuators and which will lead to further SMA applications.

5.12   Shape Memory Alloys and their Fit to Smart Technologies

It is beyond the scope of this chapter to argue deeply about the definition of what constitutes a smart material or structure. However, the following definition based on reports resulting from the UK Department of Trade and Industry's Overseas Science and Technology Expert Mission to Japan, does summarise the important issues concerning the nature of the field and the materials that can be included within it:

Smart Materials Materials with inherent functionality designed at a molecular level.

Smart Structures Systems with added functionality imparted by the integration of physical elements such as sensors and actuators with non-active materials.

It is possible to see how shape memory alloys fall into both categories. The smart material defined above requires design at a molecular level such that structural integrity, sensor and actuator functions are combined at a microscopic level to form a monolithic material rather than the macroscopic level where monolithic elements are combined to form a smart structure.

5.12.1   *Shape Memory Alloys — A Smart Material?*

The concept of smart materials and their design at a molecular level is usually interpreted as the manipulation and integration of discrete molecules, each one of which has either: a sensor, actuator or structural function. In this respect a smart material is actually a smart structure on a microscopic level. In shape memory Alloys these three functions are combined at an atomic level. In this respect the ability of shape memory alloys to respond autonomously to external stimuli is inherent and therefore fulfils one of the most important requirements of a smart material; although with relatively limited range of functionality.

In addition, a smart material should be able to respond in a controllable manner that is pre-programmed during its manufacture and processing. Again this is easily achievable with SMAs. Careful control of the alloy content and subsequent processing can be used to 'fine tune' the transformation temperatures (sensor function), the force and degree of shape recovery (actuator function) and the strength of the alloy (structural function).

Another requirement of smart materials is that actuation should only occur where it is needed. Actuators dispersed at a molecular level within a monolithic structure would on the face of it seem very difficult. However, a new method of setting the memory into discrete areas using a laser means that monolithic shape memory devices can now be made with actuators dispersed within an otherwise entirely structural matrix. A microgripper using this technique has been commercially produced and may in this respect be one of the first truly "smart material" applications of SMAs [44].

5.12.2 *Shape Memory Alloys in Smart Structures*

Essentially, two approaches have been adopted in the use of SMAs within smart structures. The first is to use SMA actuators in a relatively conventional form to drive the structure of interest. This is essentially an extension of conventional mechatronics, with complex actuation systems replaced by shape memory alloys. However, such applications differ from more conventional SMA applications since actuators are driven electrically, allowing integration of the actuation system with the control and sensing parts of the smart structure. Within this family of applications are those classed simply as "smart mechanisms". Examples include variable geometry aerofoils/hydrofoils which have been developed as demonstrators by a number of centres [45]. In these "mechanisms" SMA actuators are used to vary the geometry of 'flexible' structures in a manner analogous to human muscles, by using the varying strain available from an SM actuator as it is heated proportionately through its transformation/reversion ranges, and the mechanics of actuation are relatively conventional and simple.

At the other extreme of 'conventional' actuation is the use of SMAs in more complex structural systems. Such an example is the development of active struts for truss structures. Lightweight space truss-structures require active control to eliminate vibration and resonance. The incorporation of active struts, which can excite a structure with a cyclically imposed strain, can produce damping. A major problem is usually low frequency vibration, and low-bandwidth high-strain actuators are optimal for such active trusses. SMAs fulfil this requirement and designs have been developed for trusses based on SMAs [46]. These are relatively conventional in the sense that they use the SMA in a partially constrained actuation mode, similar to many conventional thermal sensor applications, but with the active trusses driven electrically.

A second approach to SMA-based smart structures is a more radical departure from conventional applications. This is to use SMAs as distributed strain actuators either embedded within or surface mounted to structures. This is currently the most widely investigated area since SMA wire actuators integrate well with advanced composite materials. The use of embedded SMA actuators has been demonstrated for vibration and shape control of composite components, and preliminary work has shown the potential of such actuators for damage control.

5.12.2.1 *Passive Composite Structures*

These rely on the inherent damping capacity of SMA alloys in the martensitic phase and do not involve any macroscopic shape change or motion from the 'actuator'. An example of this is the alpine ski developed at EPFL in Lausanne, Switzerland, where CuZnAl SMA alloys are integrated within a modified ski to damp vibration [47]. The damping mechanism relies on very small movements between martensite variants within the SMA to absorb vibrational energy through the hysteresis associated with reversible mechanical reorientation of a martensitic material. As the ambient temperature of the skiing conditions falls and the snow surface becomes harder, the martensitic transformation increases the damping capacity of the ski. A similar strategy is also employed for the damping of seismic vibrations in building structures using superelastic NiTi alloys [48].

Unfortunately passive control is limited to specific operating conditions. There is therefore more interest in the *active control* of smart structures as this can result in a much greater operational envelope.

5.12.2.2 *Structural Shape Control*

Many workers have demonstrated structural shape control by incorporating SMA actuators away from the neutral axes of composite components; bending introduced by actuators controlling the component's 'shape'. If actuators are placed either side of the neutral axis they can also work against one-another to provide sensitive control of shape deformation. Two additional forms of shape control have also been demonstrated which do not per se depend on the macroscopic strains of distributed actuators. These have been termed active property tuning (APT) and active strain energy tuning (ASET).

APT involves no shape change in an actuator but instead depends on its stiffness change when heated through the transformation range; transformation from martensite to parent-phase increasing the effective stiffness of a SMA by more than 200%.

Such changes can influence the stiffness of the host composite materials resulting in control of their deflection under imposed loads. For example, modelling has shown that in quasi-isotropic carbon fibre reinforced plastics (CFRP), the deflection of clamped simply supported plates can be reduced by 6% using APT; although the volume fraction of SMA required is high ($\sim$ 50%). The actuator must however be oriented correctly to maximise this effect.

During ASET an actuation strain is induced in the composite, however, no macroscopic shape change occurs since actuators are distributed throughout the host composite. This introduces internal stress into the hybrid material. Models exploring this effect have shown that the deflections of loaded simply supported plates can be reduced, and that a smaller volume fraction of actuator is required compared to APT, to produce control over deflection. For example, in quasi-isotropic CFRP laminates $\sim$ 80% reductions in deflection can be achieved with only 10 vol% actuator. The reduction of deflection also depends on the recovery force of the actuator (i.e. the amount of memory strain).

A good example of active shape control is the work carried out at Cranfield University, shown in Fig. 5.9. Here it is used to change the 'angle of attack' of a glass reinforced plastic wing structure. By embedding just two SMA wires at the root of a composite wing, (at an angle of 45° to the major axis of the aerofoil), its angle of attack can be actively controlled in real-time.

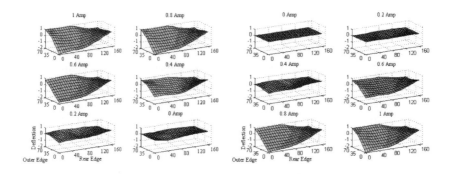

Fig. 5.9   Active shape control of a model adaptive wing.

These examples illustrate the effects of not only varying level of actuation within individual SMA actuators, but also the possibility of activating actuators at different positions within the composites and in different volume fractions, to produce widely variable actuation/shape responses.

5.12.2.3   *Vibration Control*

The second major area of interest in the application of SMA smart composites is in vibration control. In smart materials vibration control has been a major area of investigation with particular emphasis on the use of piezoelectric materials as high band-width, low-strain actuators. However, there are applications where lower band-width high strain-output actuators are required (e.g. for low frequency vibration control) and it is in this area that one strategy employing SMAs has been developed. Vibration control in smart structures has traditionally involved the sensing of vibration (or strain) within a structure, and the imposition of 'anti-strain' through a closed loop control system. Such a use of SMAs essentially involves the design of a suitable actuator and its control strategy, and work is well developed in a number of applications. However, there are other possible uses of SMAs, most again associated with their use in advanced composites.

Conventional closed-loop feedback systems between strain sensors and actuators can be used to create smart structural systems, but in no way create truly adaptive materials. However, the incorporation of SMAs into composite materials can produce such a functional composite. For example, one whose resonance frequencies can be varied by means of an external control variable. The approach here is not to strain-follow (to produce vibration cancellation), as in the case of piezoelectric actuators, but to modify composite properties such that resonance frequencies are actively moved away from excitations present within a structure. It has already been shown that effective bending stiffness can be modified by both APT and ASET, and both approaches can also be used to control resonance frequencies. APT can significantly modify the lower resonance modes of isotropic CFRP plates; shifts of resonance to higher frequencies depending on both the degree of activation (i.e. the effective stiffness change), the orientation of the actuators, and volume fraction of actuator. SMA composites controlled by APT not only result in shifts in resonance frequency, but can also control mode shapes.

It will also be of no surprise that ASET can also control resonance frequencies and mode-shapes. Doublings of the first natural frequency have been reported using only 10 vol% SMA actuator incorporated in quasi-isotropic CFRP laminates and ASET also directly modifies mode shapes.

5.12.2.4 *Buckling Control*

APT and ASET introduce in-plane loads into composite structures which not only modify their apparent flexural stiffness and modal responses, but also alter buckling characteristics. For example, APT has been shown [49] to increase buckling loads as a result of the NiTi's higher Young's Modulus in the parent-phase. Modelling of ASET has also shown a wide range of control options which depend on the orientation of the embedded actuators (compared to that of the applied load).

In the case of buckling, some components more typical of aerospace structures have also been considered. Finite element analyses of buckling blade and T-stiffened panels have been carried out with and without NiTi actuators providing control. Improved buckling [50] loads have been predicted for 1st to 5th buckling modes, and show that at high actuation strains, up to a 12% increase in load can be achieved for T-stiffened panels and $\sim 4\%$ for the blade stiffened. If multiple actuators are placed in a blade stiffened panel, increases in buckling load can also be raised further.

5.12.2.5 *Acoustic Radiation*

In addition to the control of shape, vibration and buckling, APT and ASET, have also shown potential for the control of acoustic transmission and radiation efficiency [49; 43]. Acoustically excited SMA reinforced composites can adaptively change radiation efficiency, transmission loss and directivity patterns for transmitted sound. The main processes exploited in such control have been to shift minima in transmission loss spectra, as well as changing the radiation efficiencies associated with mode shapes. The latter provides much promise for the control of sound radiation over a broad band of frequencies and the former the possibility of 'tuning' acoustic panels.

5.12.2.6   *Active Damage Control*

The final area of current interest is the use of SMA actuators for active damage control. This uses embedded sensors to detect the presence of cracks or other damage within a structure and SMA actuators to prevent propagation or damage becoming critical. Two main approaches have been investigated. In the first prememorised SMA wires embedded within composite materials are activated by heating. They attempt to return to their memorised length and result in a change in the internal stress–strain distribution. This approach has been analysed [52; 53] and shows that the recovery force under these conditions acts as a concentrated force at the free edge of the SMA hybrid composites. Because of the low internal stresses generated, this approach is not very effective [54] in influencing the stress-intensity at a crack-tip unless the SMA actuator bridges a crack. In such circumstances the crack surface then becomes a free surface, and the actuator imparts a normal stress at the tip leading to closure and a reduction in stress-intensity.

In a second concept, SMA wires with no memory are embedded into a composite component. When a crack propagates and is close to or passes through the embedded SMA actuators its stress-field results in a large strain in the SMA actuators. This effectively primes the one-way memory effect around the crack-tip. When these actuators are subsequently activated, a recovery stress is generated only in the vicinity of the crack-tip, causing reduction in stress-intensity. This approach relies on the deformation of the matrix around the actuators since deformation is required to induce the one-way effect. It has therefore been speculated [55] that this mechanism will be much more appropriate for higher strain thermoplastic, rather than brittle thermoset polymer matrices.

5.13   Final Thoughts

This chapter has examined some of the issues associated with the fit of shape memory alloys within smart technologies. The differentiating benefits of certain shape memory properties offer real opportunities for innovation. If this potential is ever going to be realised however, consideration should be given to how these market opportunities can best be exploited.

Both reactive market opportunities and proactive technological opportunities need to be present for successful innovation through shape memory alloys. It is interesting that the two markets exerting a pull on shape memory alloys are essentially concerned with utilising different memory effects. That is, the major applications and patents within the medical sector are concerned with superelasticity, whilst the smart structures researcher's concern is that of repeatable shape change in actuator applications.

Medical device manufacturers and smart structures researchers are requiring higher performance and more functional materials and devices. Developments in these areas are already exerting a pull on shape memory alloy development and this is likely to intensify as the need for smart technologies increases.

Bibliography

[1]   Perkins, J. (1974) *Mat. Sci. & Eng.*, **51**, 182.

[2]   Schroeder, T. A. and Wayman, C. M. (1977) *Scr. Metal.*, **11**, 225.

[3]   Saburi, T. and Nenno, S. (1974) *Scripta Met.*, **8**, 1363.

[4]   Oshima, R. and Naya, E. (1975) *J. Japan. Inst. Met.*, **39**, 175.

[5]   Takezawa, T. and Sato, S. (1976) *Proc. 1st JIM Int. Symp. on New Aspects of Martensitic Transformations*, pp. 233.

[6]   Zhu and Yang (1988) *Scripta Met.*, **22**, 5.

[7]   Ölander, A. (1932) *Z. Krystall*, **83A**, 145.

[8]   Greninger, A. B. and Mooradian, V. G. (1938) *Trans. AIME*, **128**, 337.

[9]   Chang, L. C. and Read, T. A. (1951) *Trans. AIME*, **189**, 47.

[10] Buehler, W. J. and Wiley, R. C. (1965) US Patent 3 174 851.

[11] Duerig, T. W. (1990) *Engineering Aspects of Shape Memory Alloys*, Butterworth–Heinemann.

[12] Wever, D. J., Veldhuizen, A. G., Sanders, M. M., Schakenraad, J. M. and van Horn, J. R. (1997) *Biomat.*, **18**, 1115.

[13] Dutta, R. S., Madangopal, K., Gadiyar, H. S. and Banerjee S. (1993) *Brit. Corr. J.*, **28**, 217.

[14] Ryhanen, J. (1999) PhD thesis, Department of Surgery, Oulu University, Finland.

[15] Fernald, R., Fritz, D., Sievert, C. and Stice, J. (1994) *Proc. 1st Int. Conf. on Shape Memory and Superelastic Technologies, Pacific Grove, California*, pp. 341.

[16] Finander, B. V. and Liu, Y. (1994) *Proc. 1st Int. Conf. on Shape Memory and Superelastic Technologies, Pacific Grove, California*, pp. 151.

[17] Andraesen, G. F. (1980) *Am. J. Orthod.*, **78**, 528.

[18] Shape Memory Applications Inc., USA, (1996) *Shape Your Future with...*, product brochure.

[19] Duerig, T., Pelton, A. and Stockel, D. (1999) *Mat. Sci. & Eng. A*, **273–275**, 149.

[20] Cwikiel, W., Stridbeck, H. and Trandberg, K. G. (1993) *Radiology*, **187**, 661.

[21] Furukawa Electric Co. Ltd., Japan, product brochure.

[22] Molloy, K. (1995) *The Mail on Sunday*, 28th May 1995, pp. 42.

[23] Thomas Bolton Ltd., UK, *SMA The Metal With a Mind*, product brochure.

[24] Kapgan, M. and Melton, K. N. (1990) *Engineering Aspects of Shape Memory Alloys*, Butterworth–Heinemann.

[25] Raychem, USA, *Unilok*™ *Rings*, product brochure.

[26] Harrison, J. D. and Hodgson, D. E. (1975) *Shape Memory Effects in Alloys* (ed. J. Perkins), Plenum Press, pp. 517.

[27] Michael, A. D. (1994) *Proc. 1ˢᵗ Int. Conf. on Shape Memory and Superelastic Technologies, Pacific Grove, California*, pp. 283.

[28] van Moorlegham, W. and Otte (1990) *Engineering Aspects of Shape Memory Alloys*, Butterworth–Heinemann.

[29] Kao, M., Schmitz, D., Thoma, P., Klaus, M. and Angst, D. (1996) *Proc. 5th Int. Conf. on New Actuators — Actuator 96, Bremen, Germany*, pp. 124.

[30] Hashimoto, M., Takeda, M., Sagawa, H., Chiba, I. and Sato, K. (1985) *J. Robotic Sys.*, **2**, 3.

[31] Guénin, G. and Gaudez, P. (1996) *Proc. 3ʳᵈ Euro. Conf. on Smart Structures and Materials and Int. Conf. on Intelligent Materials, Lyon, France*, pp. 493.

[32] Benzaoui, H., Lexcellent, C., Chaillet, N., Lang, B. and Bourjault, A. (1997) *J. Intell. Mat. Sys. & Struct.*, **8**, 619.

[33] Tamura, H., Suzuki, Y. and Todoroki, T. (1986) *Proc. Int. Conf. on Martensitic Transformations, Nara, Japan*, pp. 736.

[34] Melton, K. N. and Harrison, J. D. (1994) *Proc. 1ˢᵗ Int. Conf. on Shape Memory and Superelastic Technologies, Pacific Grove, California*, pp. 187.

[35] Buehler, W. J., Gilfrich, J. W. and Wiley, R. C. (1963) *J. Appl. Phys.*, **34**, 1475.

[36] Buehler, W. J. and Wiley, R. C. (1965) US Patent 3 174 851.

[37] Myers, S. and Marquis, D. G. (1969) *Successful Industrial Innovation*, National Science Foundation.

[38] Beta Metals, company brochure.

[39] Stoeckel, D. (1990) *Adv. Mat. & Processes*, Oct. 1990, 33.

[40] Thompson, B. S. and Gandhi, M. V. (1990) *Smart Materials and Structure Technologies: The Impending Revolution*, Technomic.

[41] Chamberlain, G. (1995) *Design News*, **50**, 70.

[42] Sheehan, T. (1995) *The Architects Journal*, **202**, 37.

[43] Rogers, C. A. (1995) *Scientific American*, Sep. 1995, 122.

[44] Bellouard, Y., Lehnert, T., Bidaux, J. E., Sidler, T., Clavel, R. and Gotthardt, R. (1999) *Mat. Sci. & Eng. A*, **273–275**, 795.

[45] Wield, D. N. and Gillam, E. (1972) *Scripta Met.*, **6**, 1157.

[46] Maclean, B. J., Draper, J. L. and Misra, M. S. (1991) *Proc. 1ˢᵗ Joint US–Japan Conf. on Adaptive Structures, Hawaii*, pp. 1038.

[47]   Gotthardt, R., Scherrer, P. and Stalmans, R. (1999) *Proc. Int. Symp. on Shape-Memory Materials, Kanazawa, Japan*, pp. 475.

[48]   Casciati, F., Faravelli, L., and Petrini L. (1998) *Proc. 4th Euro. Conf. on Smart Structures and Materials and 2nd Int. Conf. on Micromechanics, Intelligent Materials and Robotics, Harrogate, UK*, pp. 321.

[49]   Liang, C., Jia, J. and Rogers, C. (1989) *Proc. 30th AIAA/ASME/ASCE/AHS/ASC Structures, Structural Dynamics and Materials Conf.*, pp. 1504.

[50]   Thompson, D. H. and Griffin, O. H. (1992) *Proc. Conf. Recent Advances in Adaptive and Sensory Materials and their Applications, Blacksburg, Virginia*, pp. 377.

[51]   Rogers, C. A., Fuller, C. R. and Liang, C. J. (1990) *J. Sound & Vib.*, **136**, 164.

[52]   Dasgupta, A., Wan, Y. and Sirkis, J. S. (1992) *Smart Mat. & Struct.*, **1**, 101.

[53]   Rogers, C. A., personal communication.

[54]   Jensen, D. W., Pascaul, J. and August, J. A. (1992) *Smart Mat. & Struct.*, **1**, 31.

Chapter 6

Piezoelectric Materials

Tim King[1] and Michele Pozzi[2]
[1]School of Mechanical Engineering,
University of Leeds,
Leeds LS2 9JT, UK.
[2]Sirindhorn International Institute of Technology,
Thammasat University,
Pathum Thani 12121, Thailand.

6.1 Introduction to Piezoelectricity

The phenomenon of piezoelectricity has been known for more than a century. In fact, in 1880 Jacques and Pierre Curie reported to the Académie de Sciences about a new discovery: suitable compression of a hemihedral* crystal with oblique faces leads to the development of electrical charges on the surfaces. Such crystals were already known for the phenomenon of pyroelectricity†: electrical charges develop on the surfaces normal to the hemihedral axis following heating or cooling. The Curie brothers realized that the underlying explanation was identical: the variation in temperature causes a dimensional change, which is ultimately responsible for the electrical charges.

---

*A crystal is said to be hemihedral if it has at least one axis whose ends are dissimilar. From a symmetry point of view, the crystal lacks a centre of symmetry.

†The term pyroelectricity comes from Ancient Greek and describes the fact that heat ($\pi \upsilon \rho$ = fire) produces electricity. In the same way piezoelectricity is pressure ($\pi \iota \varepsilon \zeta \varepsilon \iota \nu$ = pressure) that induces electricity.

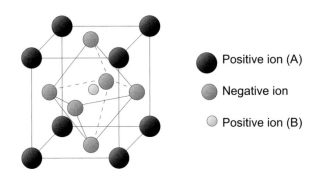

Positive ion (A)

Negative ion

Positive ion (B)

Fig. 6.1   In the elementary cell of some crystals, the 'centres of gravity' of positive and negative charges do not coincide: an electric dipole is created.

6.1.1   *Crystallography of Piezoelectricity*

Piezoelectricity can be observed in some crystals which lack a centre of symmetry. If the elementary unit (or 'cell') of the crystal lattice is such that the 'centre of gravity' of its positive charges does not coincide with that of its negative charges it creates a permanent 'dipole'[‡] — the cell is electrically asymmetric as illustrated in Fig. 6.1. A macroscopic electrical polarization is observed if the dipoles are aligned throughout the crystal. Applying external mechanical stress will strain the dipoles which alters the polarization so that electrical charges appear on the surface of the crystal. This is termed the *direct piezoelectric effect* (Fig. 6.2(a)). Conversely, applying an external electric field to the crystal will deform the natural dipoles inducing strains which change the dimensions of the crystal. We call this the *inverse* (or *converse*) *piezoelectric effect* (Fig. 6.2(b)).

Piezoelectricity is exhibited by many naturally occurring materials: quartz is an important example and tourmaline, lithium sulphate and potassium tartrate are also significant. In such crystals, however, the piezoelectric effects are not always very strong.

Ferroelectrics are an important subset of piezoelectric materials which have permanent electric dipoles in their lattice cells which can be 'switched' or re-oriented under certain circumstances. At high temperature, these

[‡]Piezoelectric behaviour can be still be exhibited in the absence of a permanent dipole if a dipole is created by inducing a separation between the positive and negative charge centres through the application of an external electric field or mechanical stress. This is the case for quartz crystals for example.

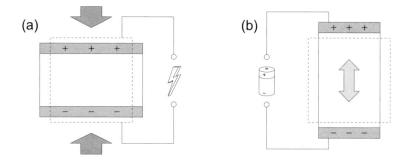

Fig. 6.2 Exemplification of (a) the direct piezoelectric effect and (b) the converse piezoelectric effect.

crystals have high symmetry, but on transition through their specific *Curie temperature* they undergo a crystallographic transformation. The low-temperature crystal structure (ferroelectric) has a symmetry which is a subgroup of that found at high temperature (paraelectric). It is worth noting that the terminology is analogous to, and derived from, that used to describe magnetic properties — e.g. ferromagnetic, paramagnetic, etc.

A historically important example of ferroelectric material is barium titanate ($BaTiO_3$), which shares the Perovskite (calcium titanate) crystal structure. At high temperature it has a perfectly centrosymmetric cubic structure: the oxygen atoms are in tetrahedral coordination with vertices sharing; at the centre of the tetrahedron is Ti (see Fig. 6.3(a)). When the temperature is lowered below 120°C (its Curie temperature), the symmetry is reduced to tetragonal. What happens is that the unit cell becomes stretched along one axis and the positive and negative atoms in it move in opposite directions — the most noticeable movement is by the titanium atom which approaches one of the oxygen atoms in the direction of stretching (Fig. 6.3(b)). This relative displacement of ions of opposite sign induces an electric dipole in the lattice cell.

Several other important materials undergo this ferroelectric transformation. A key factor is that a long-range correlation exists among the electric dipoles, so that macroscopic portions of the crystals (domains) show the same polarization. However, several different orientations are possible for the dipoles and normally all orientations will be present in different regions of the crystal (this is a form of *twinning*). Therefore, whilst each domain is piezoelectrically active because of the intrinsic polarization, in general

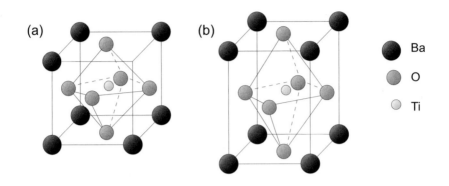

Fig. 6.3   Crystal structure of BaTiO$_3$ (a) above and (b) below its Curie temperature; the cubic cell is stretched in one direction and becomes tetragonal.

the whole crystal does not display piezoelectric effects[§]. The application of an electric field, during or after cooling, switches the dipoles and forces a common orientation (this process is called *poling*). As a consequence, the crystal as a whole becomes piezoelectrically active. Although an intrinsic polarization is present in the poled crystal, it cannot be observed directly in equilibrium conditions because electrical charges, originating either from the air or from the crystal itself, are present on its surface so as to counterbalance the internal dipole. However, the application of a mechanical stress field can alter the intrinsic polarization; before a new equilibrium is reached (by displacement or adsorption of charges), an electric charge can be observed on the surface — this is the direct piezoelectric effect.

6.1.2   *The Interaction Between Mechanical and Electrical Systems*

The uniqueness of piezoelectric materials lies in the intimate interaction between their mechanical and electrical behaviours. The strong coupling between electrical and mechanical systems implies that the mechanical properties of piezoelectric material depend heavily on the electrical parameters of the external circuit. For example, when a piezoelectric material is subjected to compressive stress, electrical charges are induced due to the direct

[§]More precisely, the whole crystal would be piezoelectrically active only if it had displayed these properties before the ferroelectric transformation, but this is not usually of practical interest.

piezoelectric effect. Looking at the phenomenon as two separate stages, one can see the induced charges as causing a strain in the opposite direction due to the inverse piezoelectric effect. The combined result is that the material appears to be stiffer because the Young's modulus receives a positive contribution from the electrical system. On the other hand, if electrodes are placed on the surface and short-circuited, the induced charges will neutralize each other and the observed stiffness will be lower, and purely mechanical. As will be shown in a following section, this can be experimentally verified by observing a shift in the natural frequency of vibration of a piezoelectric device between open- and short-circuit conditions.

6.1.3 Some Piezoelectric Materials

The Curie brothers worked on natural crystals such as tourmaline, topaz, quartz and Rochelle salt, although only the latter two are piezoelectric to a high level. Much larger piezoelectric activity is found in crystals artificially synthesised for the purpose. The already mentioned perovskite family have proved to be particularly successful: barium titanate ($BaTiO_3$), lead titanate ($PbTiO_3$ or PT) and particularly lead zirconate titanate ($Pb(Zr, Ti)O_3$ or PZT) and lead lanthanum zirconate titanate (PLZT). Similar crystal structures are found in lithium niobate ($LiNbO_3$) and lithium tantalate ($LiTaO_3$).

It is difficult to grow sizeable crystals of these materials and so they are used in ceramic form instead. Ceramics are aggregations of microcrystalline particles (*grains*), with sizes varying from several nanometres up to microns. They are usually formed by firing a compacted slurry of the finely ground constituents. This process is termed *sintering*. Because the grains, and hence their dipoles, are randomly oriented a poling process is essential before these ceramic materials can exhibit piezoelectricity.

Because sintering takes place at very high temperatures, as the ceramic cools through its Curie temperature dipoles are formed without preferential orientation within each grain. The ceramic exhibits no net dipole, and hence no piezoelectric activity. Piezoelectric properties therefore have to be developed in devices made from ferroelectric ceramics by poling. This is achieved by applying a strong electric field which orients the dipoles in each grain along the most favourable direction. Perfect alignment cannot be achieved because only specific directions are available to the dipoles within the grains whose orientations are virtually fixed during sintering, but a level of polarisation of 80% or more can be effected. Piezoelectric ceramics produced by this method have become the norm for actuator applications on account of their strongly piezoelectric properties, which can be tailored, and the flexibility to readily create a wide range of shapes.

Since the discovery of the superior properties of PZT, targeted studies have led to the development of a whole series of piezoelectric ceramics based on $Pb(Zr_{0.55} Ti_{0.45})O_3$, whose elements differ by the addition of oxides of several metals (Nb, Cr, La and Fe). This series can be divided into two main families: *soft* PZT (such as 'PZT-5H') and *hard* PZT (such as 'PZT-8'). The former are characterised by higher piezoelectric activity and therefore higher output displacements, but they also suffer from considerable hysteresis and energy dissipation. This means that they are more suitable for quasi-static actuation applications. The latter, hard PZTs, give smaller displacements for a given electric field but also very small hysteresis and energy loss, which makes them more attractive for high frequency applications.

Piezoelectricity is not limited to inorganics. It is also present in polymers having a chain structure: even the natural materials wool and silk show piezoelectricity to some extent, although significant piezoelectric activity is only found when the molecule has strong dipoles. The best example is again an artificially synthesized material: the polymer polyvinylidenedifloride (PVDF), whose monomer is $-CF_2-CH_2-$, is more piezoelectric than quartz. Although it is less active than the perovskite ceramics and can develop smaller forces, it has the advantages of being lightweight and easily shaped. In order to exhibit significant piezoelectric activity, PVDF has to

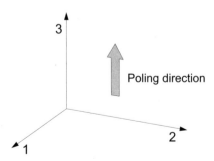

Fig. 6.4 System of axes commonly defined for piezoelectric ceramics.

be extruded and drawn so as to orient the molecules and their dipoles as much as possible. Due to the low value of its dielectric constant, the value of the piezoelectric constant g is higher in PVDF than in ceramics, making the polymer more suitable for sensing applications. It is preferable to ceramics also as an acoustic generator in medical devices because its low density and Young's modulus allow for a better acoustic coupling with the human body.

Some fundamental constants can be defined describing piezoelectric materials which allow us to compare their behaviours. For piezoceramic actuator applications a vital characteristic is the coupling between the applied electric field and the mechanical strain which results from it. We term this d. As with most materials, straining in one direction causes strains in orthogonal directions also. For piezoelectric materials it is also possible that an electric field in one direction can cause shear strains. For this reason we need to define a set of values for d relating electric field in one direction to all the possible strains which it may cause in various directions¶. To do this we first define a set of axes. By convention, for piezoceramics we align axis 3 with the poling direction (Fig. 6.4). By using subscripts we can now give names to the different d values. For example, d_{33} is the value of d relating electric field in the (poling) direction 3 to strain in the (same) 3 direction; d_{31} relates electric field in the 3 direction to strain in the (orthogonal) 1 direction. Since piezoceramic devices are almost invariably driven using the same electrodes employed for poling them, electric fields applied other than in the 3 direction will not be considered here. In practice the d_{33} and d_{31} values are the most important from the point of view of actuator applications. Values of these constants are given in Table 6.1 for sele

6.2   Applications of the Direct Piezoelectric Effect

As described in the introduction, the direct piezoelectric effect is responsible for the generation of electrical charges on a piezoelectric material subjected to mechanical stress. These charges are proportional to the intensity of the

¶In the most general case there could therefore be 18 independent d values ('piezoelectric charge constants'). However, for piezoceramic materials, symmetry considerations dictate that many of them are zero and others equivalent, which greatly reduces the number of values of interest. For example d_{32} is equal to d_{31} and hence usually not quoted. Subscripts 4, 5 and 6 relate to induced mechanical shear. Charge constants involving these subscripts are not generally significant for piezoceramic actuators.

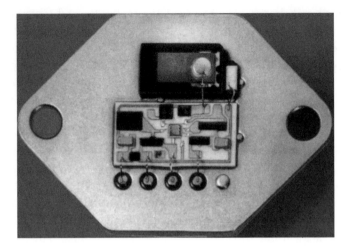

Fig. 6.5   The direct piezoelectric effect is exploited to measure accelerations. In the figure, a piezoelectric element is fitted together with the signal conditioning electronics in one single integrated circuit [12].

force applied, so that they can be used to measure the force itself. Thanks to their high resonant frequency, force sensors based on piezoelectricity have very short response times and are perfect for measuring rapidly varying forces (on the other hand they perform poorly in very low frequency applications due to drift caused by the internal electrical conduction). One of the main applications of such sensing properties is in accelerometers where a known mass translates an acceleration into a force. The technology of piezoelectric accelerometers is so well developed that they are miniaturized down to integrated circuits containing the sensing element and the signal conditioning electronics (see Fig. 6.5). cted piezoelectric materials.

The direct piezoelectric effect is also used in less high-tech applications, which are nonetheless widespread and important. For example, the high intensity of the electric field that builds up in a piezoelectric material subject to strain allows high voltages to be generated (in the order of thousands of volts). These voltages are exploited to generate low power sparks for fuel ignition applications.

An important application where the direct and converse piezoelectric effects are used together is in piezoelectric transformers. The basic idea, proposed by Rosen in the 1950s, is to excite one region of a piezoelectric material electrically with a driving sine wave and to collect the transformed

Table 6.1   Piezoelectric charge constants for several piezoelectric materials of interest.

Material	d_{33} $(\times 10^{-12}\text{mV}^{-1})$	d_{31} $(\times 10^{-12}\text{mV}^{-1})$
$BaTiO_3$	191	-78
PVDF	-33	2–14
PZT-4	289	-123
PZT-5A	374	-171
PZT-5H	593	-274

electrical output from another region which is mechanically coupled with the first [13]. These devices have higher efficiency in resonant mode than their electromagnetic counterparts and are widely used to power, for example, backlights for portable computers.

6.3   Acoustic Transducers

The first technological application of piezoelectricity was realized in the years around the First World War by Paul Langevin, who built an acoustic transducer for naval application: the first piezoelectric sonar. Thanks to the direct coupling between electrical and mechanical systems and the high resonant frequency, quartz crystals sandwiched in steel were found to be very appropriate for the emission of the sonar "chirp". Piezoelectric sonars have been improved since for better coupling with the water medium and by including piezoelectric ceramics rather than quartz. Piezoelectric materials, especially PVDF, are often used for the 'tweeters' in modern loudspeaker systems and, as mentioned before, in a host of medical applications from echo-graphic devices to ultrasound surgery.

6.4   Piezoelectric Actuators

New actuators with better performance are always sought for mechatronics devices and micro-electro-mechanical systems. Among the many technologies available, piezoelectric actuation offers high speed, precision and high energy densities. Different piezoelectric materials have been used for actuation in a wide range of applications, sometimes to exploit particular ad-

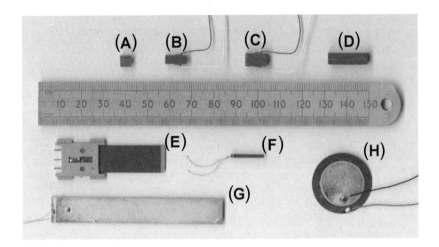

Fig. 6.6   A selection of commercial piezoelectric actuators: (A–D) multi-layer actuators; (E–G) bimorphs; (H) unimorph.

vantages; e.g. low contamination in ultra-high-vacuum environments, but more often to take advantage of the unrivalled levels of speed and precision available from piezoelectric devices. Indeed, piezoelectric actuators are sometimes essential to a whole technology — it is difficult to imagine an atomic force microscope without piezoelectric control of the scanning.

Piezoelectric actuators exploit the inverse piezoelectric effect to transform electrical energy into mechanical work. As described previously, this conversion takes place at the crystal-lattice level and is direct (by contrast, in solenoids and magnetostrictive actuators the electrical energy must first be converted into magnetic energy; or in the case of electrically driven shape memory alloys, into thermal energy). This leads to some advantages such as negligible electromagnetic noise and high energy density.

This section deals with the application of piezoelectricity to actuation technology, describing the most important configurations of piezo-actuators.

6.4.1   *Bimorphs and Other Bending Piezo-Actuators*

In its actuation principle, the piezoelectric bimorph is very similar to the more common 'bimetal strip'. In both cases a bending moment is produced by the differential contraction/expansion of two different strips joined together. While in bimetals the difference in thermal expansion coefficients is exploited, in bimorphs it is the transverse piezoelectric strain that plays this role: the coupling constant d_{31} relates the electric field parallel to the poling direction to the strain in the plane normal to it. Therefore, by applying the same electric field to two bonded strips with different poling directions, a differential contraction/expansion is produced and bending results.

Based on this principle two bimorph configurations are possible (refer to Fig. 6.7). By bonding together two layers of piezoelectric ceramics with opposite poling directions, a *series* bimorph is obtained: a voltage across the (external) electrodes of the assembly induces its bending because one layer contracts while the other expands (device F in Fig. 6.6 is a series bimorph). In the *parallel* configuration the two layers have the same poling direction: a third electrode must then be introduced between them so that it is possible to apply opposite electric fields to the layers, leading to strains of opposite signs (see device G in Fig. 6.6). In series bimorphs one of the layers is always excited with an electric field opposite to its poling direction, which results in de-poling in the long run. In the parallel configuration a biasing voltage can be applied to overcome this problem (as can be done with device E in Fig. 6.6).

Similar operational characteristics are found in *unimorphs*, where one of the active layers is substituted by a metal plate. This is the common configuration used in buzzers and sounders (one such device is shown in Fig. 6.6(H)).

Rainbow actuators are made by chemically reducing one surface of a lead-rich ceramic wafer (therefore being suitable for PZT or PLZT piezoelectric ceramics as well as to lead magnesium niobate (PMN) electrostrictive ones and others). As a consequence of the high temperature reduction, the wafer exhibits a spontaneous concavity when cooled to room temperature (*rainbow* stands for *r*educed *a*nd *i*nternally *b*iased *o*xide *w*afer). Moreover, the reduced surface is piezoelectrically inactive and represents a constraint when the unreduced and active part of the wafer is energised. For this reason rainbow actuators are somewhat similar to unimorphs, with the advantage that the mechanical constraint (and electrode) is made from the wafer itself, removing any adhesion problem.

Actuation takes place as the displacement of the mid portion of the wafer in the direction normal to its surface and is a consequence of the radial contraction/expansion of the active layer caused by the electric field through the d_{31} constant.

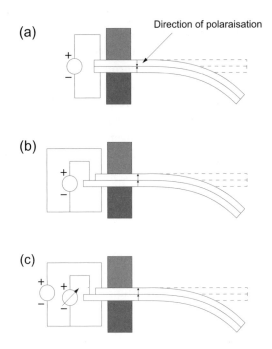

Fig. 6.7   Piezoelectric bimorphs: (a) series bimorph; (b) parallel bimorph; (c) parallel bimorph with biasing voltage.

Rainbow actuators are capable of larger displacements and higher forces than unimorph/bimorph benders and offer the additional benefit that they can be stacked to further increase the output displacement. Haertling [1] reports displacements up to 1mm and 100N forces with 0.2mm-thick wafers (see Fig. 6.8).

6.4.2   *Monolithic Actuators*

The use of single crystal piezoelectric devices has the advantage that the complexity of the electro-mechanical coupling can be fully exploited. By suitably cutting a piezoelectric crystal it is possible to obtain extension, bending or torsion through the application of an electric field. Many small displacement actuators have been realized in this way, especially for micro-positioning. To obtain somewhat larger displacements solid piezoceramic devices are also attractive. For example, a common design for sample holders is the piezoelectric tube, which is made of ceramic and longitudinally divided into four sections with separate electrodes on each quadrant. By energising the correct combinations of such electrodes, it is possible to tilt and translate the sample with great flexibility.

Since the electric fields required to achieve significant displacements are in the order of $1MVm^{-1}$, high driving voltages are usually necessary for these bulk piezoelectric devices. For pushing actuator applications this limitation is now commonly overcome by dividing the actuator into thin layers to give *stack* or *multi-layer* devices (described later). However, in special configurations, monolithic (i.e. not layered) actuators can still be very appealing. This is the case for the *Moonie* and its derivative, the *Cymbal*.

6.4.2.1 *Moonies and Cymbals*

The concept underlying the Moonie actuator was first employed in the acoustical transducer for oceanographic application patented in 1991 by Newnham *et al.* and presented later in a paper [2]. The basic idea is to apply metal end-caps to both the faces of a piezoelectric disk (as shown in Fig. 6.9(a)) so that the compressive forces orthogonal to the disc produce a radial extension of the disc as well as compression. In such a configuration, the effective d_{33} constant which relates the electric field to the strain, in this

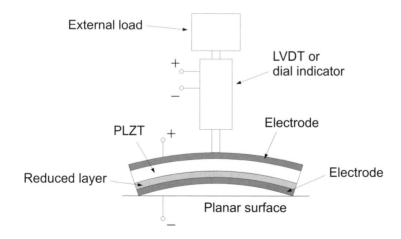

Fig. 6.8 A rainbow actuator with an LVDT displacement meter [12].

case both orthogonal to the disc direction, receives a positive contribution from the d_{31} constant of the piezoelectric disk (which relates to the strain developed in the plane of the disk by the orthogonal applied electric field). It can be shown that the following formula holds [2]:

$$d_{33}^{\text{eff}} = K_1 d_{33} + K_2 d_{31}, \qquad (6.1)$$

where K_1 and K_2 are both positive and depend on the geometric dimensions (and material characteristics) of the disc and end-caps. It is observed that large cavity diameters lead to higher values of d_{33}^{eff}, but also to lower resonant frequencies [2].

A few years later, the same researchers exploited this idea for actuating purposes, which led to the Moonie actuator sketched in Fig. 6.9(b) (Newnham *et al.* [3]). Moonie actuators are preferably made of PZT-5A discs with brass end-caps; the dimensions of the composite are about 1cm in diameter and 2 to 7mm in thickness. Effective d_{33} constants can be as high as $3 \times 10^{-9} \text{mV}^{-1}$ (later improved up to $10 \times 10^{-9} \text{mV}^{-1}$) which is equivalent to an amplification ratio of 5. It is found that increasing end-cap thickness leads to an increased resonant frequency, at the expense of a reduced displacement (and effective d_{33}). On the other hand, a reduced end-cap thickness leads to smaller output forces, although 20N can still be delivered with 0.4mm-thick end-caps. Various studies have been carried out since the first appearance of Moonies to optimise their performance. The most important improvement requires cutting a groove in the metal caps, close to the region where they are bonded to the piezoelectric disc. Since at that point there is a concentration of stress and the grove locally reduces the rigidity of the structure, a larger displacement is achieved and less energy is elastically stored in the end-caps [4]. However, the most significant improvement to the Moonie actuator came from its inventors a few years later [5] with the Cymbal design (see Fig. 6.9(c)). The major disadvantages of the previous design were the dependence of the output displacement on the contact position (the end-caps are not flat in the deformed state) and complex manufacture because of the cavity that has to be machined into the end-caps. As has already been mentioned, one suggested enhancement of the Moonie requires machining an annular groove above the bonding region ('grooved Moonie'), but this further increases production complexity. By contrast, the Cymbal actuator has punched end-caps which are flat in the middle, therefore simplifying manufacture and offering a more reliable

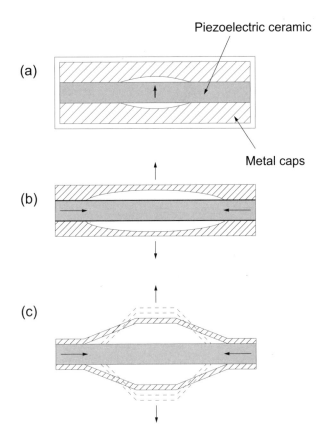

Fig. 6.9 Moonies and Cymbals: (a) Moonie acoustical transducer; (b) Moonie actuator; (c) Cymbal actuator [6].

contact surface. At the same time the thickness of the end-caps is reduced in the bonding area, which increases the flexibility and minimises mechanical losses — without the need for machining the groove. So it is found that Cymbal actuators have better performance than both the original Moonie and the 'grooved Moonie', with effective d_{33} of $15 \times 10^{-9} \mathrm{mV}^{-1}$ [5]. In addition, the flatness of the end-caps simplifies the stacking of many actuators to build an array (for greater displacement and/or output forces).

An 'exhaustive' approach has recently been developed by Silva and Kikuchi to find the best topology for a composite actuator made of a piezoelectric ceramic and some inactive material — typically metal [1]. In this

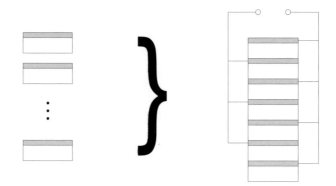

Fig. 6.10   A piezoelectric stack actuator is formed by assembling together several electroded piezoelectric elements.

method the traditional intuitive approach based on "engineering common sense" is replaced by a systematic optimisation of a structure. In the first place, an array of cells is created and allowed to have a degree of filling of a defined material. The distribution of material is then optimised either for maximum force or displacement. The algorithm developed determines automatically which cells are to be voids and which must be filled with material. Interestingly, the procedure has been applied to a Moonie-like actuator and the resulting configuration is very similar to a Cymbal.

6.4.3   *Stack and Multi-Layer Actuators*

As has already been briefly mentioned, the necessity for high driving voltages hinders the extensive application of long extensional piezoelectric actuators. This problem was ameliorated with the development of stack actuators. In this configuration, several layers of thin piezoelectric ceramics are individually provided with electrodes and then stacked one upon the other as illustrated in Fig. 6.10. The resulting structure is a succession of ceramic-electrode elements. By connecting together every other electrode, the piezoelectric elements are electrically connected in parallel, whilst mechanically they are placed in series. The obvious advantage is that the driving voltage does not act across the whole length of the actuator but only across each thin layer. In this way the resulting electric field is much higher.

A further improvement came with the multi-layer technology, initially developed for capacitors. Here the piezoelectric material is laid down in thin layers on which electrodes are deposited by screen printing. These are then stacked one upon the other, and the ceramic actuator formed by firing the assembly. The structure is illustrated in Fig. 6.11, whereas a range of real devices is shown in Fig. 6.6(A–D). This technique decreases production costs while increasing reliability. After firing the actuator is poled (in multi-layers, as in stacks, adjacent layers have opposite poling directions).

6.4.3.1 *Multi-Layer Characteristics*

Multi-layer actuators are the most robust configuration available for piezo-actuators. They are used as pushers because in this mode they can develop very high forces, whilst being rather fragile in tension, bending or torsion. The operating voltages range from about 60 to 150V or more, depending primarily on the thickness of the layers and secondarily on the material. Recently, very thin layers have been realized (down to 30µm). While this reduces the voltage necessary it also prejudices the energy density (the ratio of electrode to active material increases). The production of very thin layers requires a sophisticated approach to avoid porosity; voids in the structure can result in cracking or arcing between the electrodes.

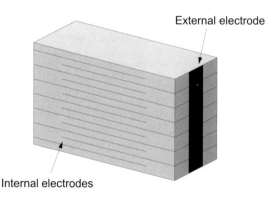

Fig. 6.11 The multi-layer structure of piezoelectric actuators: many sandwiches of electrode–ceramic–electrode are built together so that the various piezoelectric layers are electrically connected in parallel and mechanically in series.

The maximum pushing force F_0 exerted by a multi-layer is found at zero displacement and is therefore referred to as "clamping force" (or synonymously "blocking" or "stall" force). The available pushing force F decreases linearly to zero as the output displacement l tends towards its maximum value l_0:

$$F = F_0 \left(1 - \frac{l}{l_0}\right) \quad \text{for} \quad 0 \leq l \leq l_0. \tag{6.2}$$

The clamping force is mainly determined by the cross-section A of the actuator; in fact it is better expressed as a maximum compressive stress σ_0 (or pressure), so that

$$F_0 = A\sigma_0. \tag{6.3}$$

Typical values of σ_0 are around 35–50MPa leading to forces from hundreds to thousands of Newtons. If the Young's modulus Y is given in the manufacturer's datasheet together with the cross-sectional area A and the length L, the elastic constant k of the actuator can be found as

$$k = Y\frac{A}{L}, \tag{6.4}$$

and a different expression for F can be derived:

$$F = k(l_0 - l) \quad \text{for} \quad 0 \leq l \leq l_0, \tag{6.5}$$

where, once again, l_0 is the free displacement at the applied voltage.

6.4.3.2 *Dynamic Characteristics of Multi-Layers*

Data on the dynamic response of a piezoelectric actuator are not commonly found in manufacturers' datasheets. The dynamic characteristics can be evaluated by exciting the piezo-actuator with a quickly rising voltage and observing the evolution of the displacement with time. In performing the test, care is needed to limit the inrush of current which may overstress the internal electrodes and also cause high tensile stress in the actuator. A simple test method involves the use of an electronic circuit having adjustable resistive and inductive impedance and featuring an electronic switch to turn on the driving voltage.

There are several parameters of a piezo-actuator that influence its dynamic response, in addition to those of its driving circuit. One is clearly the mechanical resonant frequency, which is in turn mainly determined by

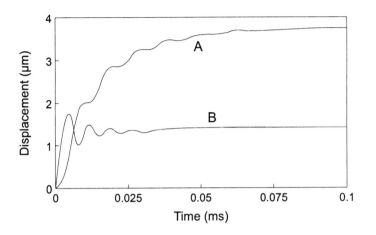

Fig. 6.12   Dynamic response of two piezoelectric actuators. Actuator A is $2 \times 3 \times 10\text{mm}^3$ and Actuator B is $5 \times 5 \times 18\text{mm}^3$. The driving circuit has negligible inductance but a resistance of 5Ω (in series with the actuator).

the Young's modulus and the length of the actuator. Electrical parameters are also important: the capacitance of the piezoelectric actuator is a factor in determining the time constant of the circuit (in conjunction with the resistance of the electrodes and any external circuit resistance) and so the time before the electrodes of the actuator reach the nominal voltage. Further, as explained in the introduction, the actual resonant frequency of the actuator is strongly affected by the piezoelectric effect and therefore by the relevant constants.

Figure 6.12 shows the response to a voltage step as observed on two different commercial actuators in the same experimental conditions: one is $2 \times 3\text{mm}^2$ in cross-section, the other is $5 \times 5\text{mm}^2$; their respective lengths are 10 and 18mm. It is possible to observe the effects of the different resonant frequencies and capacitances on the response.

Prediction of the behaviour of a piezo-actuator in a given condition is possible through a computational model that takes into due account both the mechanical and the electrical systems and their strong interaction. Such a model also needs to implement a distributed mass scheme for the actuator to be able to describe various normal modes of vibration.

It has been anticipated in the introduction that the resonant frequency of a piezoelectric material is dependent on electrical parameters. In order

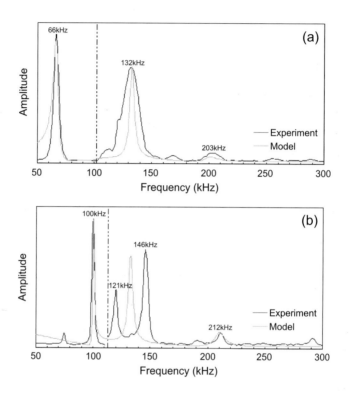

Fig. 6.13 Model and experimentally measured frequency responses for a piezoelectric multi-layer actuator. The source of excitation is mechanical, the actuator is mechanically unconstrained, but under electrically (a) open- or (b) short-circuit conditions. Amplitudes at high frequencies have been magnified for clarity.

to demonstrate this experimentally, it is possible to excite the mechanical vibration of the material either by a voltage pulse or a mechanical stimulus. Figure 6.13 shows the frequency response of a mechanically unconstrained piezoelectric multi-layer mechanically stimulated. As can be seen, resonant frequencies are affected by the electrical circuit conditions.

The phenomenon is even more interesting in a mechanically unconstrained multi-layer actuator (as in the figure) because the layers are electrically interconnected. When the multi-layer is vibrating, some sections of it may be vibrating in counter-phase to others. Referring to Fig. 6.14, one can see that in the first mode, all strain in the actuator has the same sign at a given time (either compression or tension). The second harmonic, on the contrary, sees one half of the actuator in tension while the other is under compression. In electrical terms, this means that while one half causes a positive voltage, the other causes a negative voltage. Because of the connections between the electrodes, these two electromotive forces cancel out and the parameters of the external circuit become irrelevant (the piezoelectric material is always short-circuited). A complete cancellation is expected only for even harmonic orders, while for odd orders the amount of shift decreases with increasing numbers because a smaller fraction of the material is unbalanced.

6.5   The Problem of Amplification

The major strengths of multi-layer piezoelectric actuators are speed, output force, energy density and precision. Some of these are directly related to the physical origin of the phenomenon: the conversion of electrical energy into mechanical is mainly due to a stretching of the crystal lattice. Naturally this deformation is limited and so only small displacements are practicable (usually less than 0.1%, i.e. in the order of tens of microns for actuators of reasonable length). This is a serious problem for applications where mechanical tolerances are not very precise or simply when larger displacements are needed. For this reason research has been concentrated both on developing new materials with enhanced performance and also on designing auxiliary devices to exploit present materials to their maximum. The amplification of the displacement from a piezo-actuator can be achieved in many different ways. However, amplifiers can be grouped into three

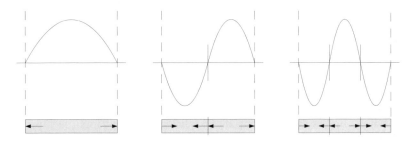

Fig. 6.14   The first three vibration modes of a multi-layer actuator. The graphs above show, at a given time, the strain as a function of the position inside the actuator. The sketches below them exemplify the states of tension/compression.

main families according to their operating principle. Mechanical amplifiers mainly exploit levers and hinges; 'integrating' amplifiers add together many small steps to achieve a larger displacement; finally, impact amplifiers exploit the fast response of a multi-layer actuator to impart kinetic energy to a projectile that can then travel larger distances. Sometimes these basic principles are combined in hybrid amplifiers with the aim of reducing some of the limitation intrinsic to each separate technique.

6.5.1   *Mechanical Amplification*

Here mechanical amplification means a direct and 'static' magnification of the output displacement of a piezoelectric element via added mechanical devices of various complexity. It is 'static' in the sense that all the parts are always in contact, without slipping or bouncing.

Mechanical amplifiers can roughly be divided into two categories: whilst simple lever amplifiers exploit the lever principle, so that the movement of the piezoelectric device (high force, small displacement) is transferred to the output of the actuator assembly with lower force and higher displacement, flextensional actuators magnify the input displacement thorough the buckling of a slender or thin element.

Some of these mechanical amplifiers require the rotation of parts around some pivots. Because of the small displacements involved, backlash is unacceptable and common bearings are replaced by flexural hinges [7] (this also gets rid of tribology problems, although it introduces new risks of fatigue failure).

Simple lever amplifiers are based on the lever principle, as illustrated in Fig. 6.15(a). High amplification factors can be obtained in theory, but some limitations must be considered. First of all, a long lever means a large mass to displace, which reduces the resonant frequency and the speed of response. Moreover, machinability and stress concentration in the hinges set some limits to the dimensions of the elements in the lever, and so to the magnification ratio. A possible means to higher amplification is a two-stage lever amplifier, possibly with the integration of light weight materials and composites. In simple lever amplifiers, efficiency is kept high because only two hinges are necessary, but the output displacement has a curvilinear character that may be undesirable in precision-oriented applications.

Flextensional actuators are those devices in which a beam or similar element is caused to bend by a compressive force exerted at its ends by one

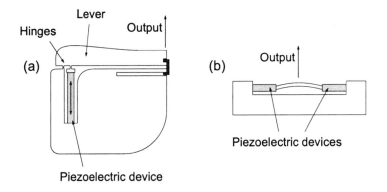

Fig. 6.15   Working principle of (a) lever amplifiers and (b) flextensional actuators.

or more piezoelectric devices (an example is reproduced in Fig. 6.15(b)). As a result the mid portion of the beam describes a straight-line motion at right-angles to the input displacement (from the piezo). Numerous different designs can be conceived based on this principle . Examples of the simplest embodiment are found in amplifiers where flexural hinges are exploited to concentrate bending in defined pivots. In other flextensional actuators there are no hinges at all and the stresses are more uniformly distributed on an elliptical shell. The geometrical parameters of the shell, like the eccentricity, can be tailored to the required output characteristics.

6.5.2   *The Summation of Multiple Small Steps*

An amplification technique which allows 'unlimited' output displacement is the integration of multiple small steps. The main embodiments of this principle are the *inchworm* motor and the inertial *stick and slip* actuators which are briefly discussed in this section along with the principle of the *ultrasonic travelling wave motor*, where a dynamic integration of a continuum motion is realised.

The principle of the inchworm motor, which has been subject of some patents, is shown in Fig. 6.16. The active elements are three piezoelectric actuators bonded together: the one in the middle (PZ2) suffers a strain in the longitudinal direction while the others (PZ1 and PZ3) alternately clamp the moving rod. While the actuator PZ1 (which is assumed to be held stationary) clamps, PZ2 expands pushing PZ3 (which is now unclamped)

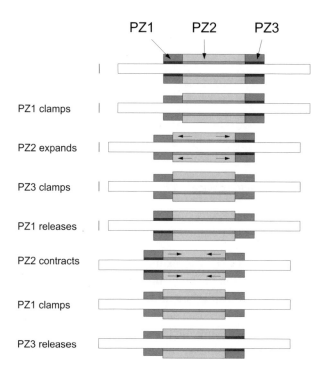

Fig. 6.16   Operation of the inchworm motor.

forward: at this stage the rod does not move (stage 1). Then PZ1 releases its grip and PZ3 clamps the rod, whereupon PZ2 contracts, drawing PZ3 and the rod towards PZ1 (stage 2). This results in a fairly smooth motion when many such steps are made, high output forces (around 10N) and high mechanical resolution (in the order of nanometers); the only limitations being the maximum velocity (in the order of $10^{-3}\mathrm{ms}^{-3}$) and the tight mechanical tolerances required in the production of the motor.

The inertial approach to the integration technique is based on the quick response of piezoelectric actuators (thus bearing some resemblances with the impact technique described in the following). A sawtooth voltage is applied to some piezoelectric elements on which a massive platform is resting (Fig. 6.17): during the linear rise, friction prevails so that the platform translates, to the right in the figure (from a to b). When the electric field is suddenly reversed, friction is overwhelmed by inertia: the platform does not move but the piezoelectric elements regain their original position, moving to the left (from b to c). The typical speed achievable is around 10^{-4}ms^{-1} over a few centimetres and with steps in the order of tens of nanometers (but sub-nanometer resolution is possible between jumps through fine control of the voltage in the linear rise), while output forces are obviously very small (tens of mN). This principle has been widely exploited for micropositioning in microscopes (e.g. STMs and SPMs), where waveforms more complex than the sawtooth have also been investigated [9].

In ultrasonic travelling wave motors two standing resonant waves are induced in an elastic thin medium with a relative phase shift so that their superimposition causes every point on the surface to move on an elliptical trajectory. If an object free to move is placed on the elastic medium, it will translate. Both circular and linear motors have been built based on this principle. Such rotary motors exhibit high torque even at low speeds, with zero power consumption in the holding position; moreover they have the advantage of being free from electromagnetic interference.

It is clear that the main advantage of the integration principle is the virtually unlimited displacement available. Unfortunately the main drawback is the reduced speed and force outputs.

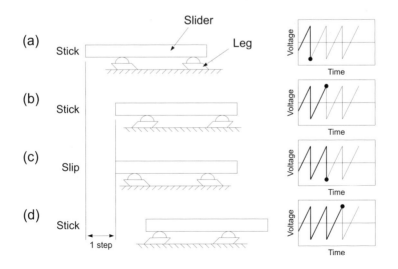

Fig. 6.17   Sketch of an inertial actuator in operation and the signal fed to it [8].

6.5.3 *The Impact Technique*

The underlying idea of the impulse actuator is the transfer of energy, in kinetic form, from the piezoelectric element to a projectile. Due to the fast response of the piezoelectric actuator this energy transfer has an impulsive character. It must be noticed that after having received the impulse, the projectile will travel in 'free flight' so that its motion becomes independent from the displacement of the piezoelectric element. Therefore in impact actuators it is not possible to have active control over the output displacement and only the initial velocity can be chosen to some extent. On the other hand, the distance travelled can be very large.

An impact actuator for dot matrix printers has been developed and patented by Chang *et al.* [10]. In this application the piezoelectric element is clamped in the middle of an E-shaped frame and kept in a compressive state so that tensile stresses are avoided (see Fig. 6.18). Thanks to its fast response, the piezoelectric actuator generates a shock wave which propagates in the central beam of the frame and is then focused by a conical wave-guide before being transmitted to a hammer. Having received a momentum, the hammer and the printing wire connected to it travel toward the paper. After the impact of the printing wire against the paper, the hammer bounces back. In its way back it is guided by a leaf spring which also ensures the contact between hammer and wave-guide between successive actuating events. Travels up to 0.38mm have been reported,

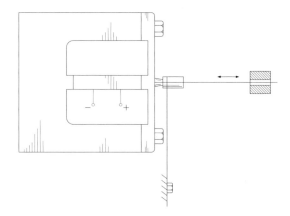

Fig. 6.18   Impulse actuator for dot matrix printing.

with initial velocities greater than $2.5ms^{-1}$ [10]. It is worth noticing that at present the exploitation of the impulse approach is essentially limited to such applications. One reason of the limited interest in this kind of amplification is probably the impossibility of having a proportional control on the actuation process, which confines it to 'discrete' (or non-proportional) actuation systems.

6.6   Further Application Examples

Piezoelectric materials, typically quartz, are used to build reliable oscillators having high frequency stability. Their applications are numerous, ranging from quartz clocks to radio transmitters and receivers. Nowadays, very large quartz crystals are grown very commonly through the hydrothermal method, which guarantees high production volumes and higher quality than natural crystals.

Bibliography

[1]   Haertling, G. H. (1994) "Rainbow Ceramics — A New Type of Ultra-High-Displacement Actuator", *Am. Ceramic Soc. Bull.*, **73**, 93–96.

[2]   Xu, Q. C., Yoshikawa, S., Belsick J. R. and Newnham R. E. (1991) "Piezoelectric Composites with High Sensitivity and High Capacitance for Use at High Pressures", *IEEE Trans. Ultrasonics, Ferroelectrics and Frequency Control*, **38**, 634–639.

[3]   Newnham, R. E., Xu, Q. C. and Yoshikawa, S. (1997) "Metal-Electroactive Ceramic Composite Actuators", US Patent 5 276 657.

[4]   Shih, W. Y., Shih, W. H. and Aksay, I. A. (1997) "Scaling Analysis for the Axial Displacement and Pressure of Flextensional Transducers", *J. Am. Ceramic Soc.*, **80**, 1073–1078.

[5]   Dogan, A., Uchino, K. and Newnham, R. E. (1997) "Composite Piezoelectric Transducer with Truncated Conical End-Caps Cymbal", *IEEE Trans. Ultrasonics, Ferroelectrics and Frequency Control*, **44**, 597–605.

[6]   Newnham, R. E. and Dogan, A. (1998) "Metal-Electroactive Ceramic Composite Transducer", US Patent 5 729 077.

[7]   Xu, W. and King, T. G. (1996) "Flexure Hinges for Piezo-Actuator Displacement Amplifiers: Flexibility, Accuracy and Stress Considerations", *Precision Eng.*, **19**, 4–10.

[8]   Breguet, J. -M. and Clavel, R. (1998) "New Designs for Long Range, High Resolution, Multi-Degrees-of-Freedom Piezoelectric Actuators", *Proc. 6th Int. Conf. on New Actuators — Actuator 98, Bremen, Germany.*

[9]   Smith, W. F., Abraham, M. C., Sloan, J. M. and Switkes, M. (1996) "Simple Retrofittable Long-Range x-y Translation System for Scanned Probe Microscopes", *Rev. Sci. Instr.*, **67**, 3599–3604.

[10] Chang, P. S. H. and Wang, H. C. (1990) "A High Speed Impact Actuator using Multilayer Piezoelectric Ceramics", *Sensors & Actuators A*, **24**, 239–244.

[11] Silva, E. C. N., Nishiwaki, S. and Kikuchi, N. (2000) "Topology Optimiza-

tion Design of Flextensional Actuators", *IEEE Trans. Ultrasonics, Ferro-electrics and Frequency Control,* **47**, 651–656.

[12] Waanders, J. W. (1991) "Piezoelectric Ceramics: Properties and Applications", *Philips Components,* ISBN: 9398 651 80011.

[13] Rosen, C. A. (1956) "Ceramic Transformers and Filters", *Proc. Electronic Components Symp.*, pp. 205–211.

Chapter 7

Magnetostriction

Alan G. Jenner[1] and Donald G. Lord[2]

[1]Department of Physical Sciences,
University of Hull,
Hull HU6 7RX, UK.

[2]Joule Laboratory, School of Science,
University of Salford,
Salford M5 4WT, UK.

7.1 Introduction

The so-called "smart materials" refers to a class of materials that are highly responsive to an outside influence (either electrical or magnetic, for example) and have the capability to sense and react to changes within themselves. Smart materials have a significant number of uses in commercial as well as military applications. The worldwide sales of smart materials exceed $1 billion annually.

The choice of smart material for an application depends on the nature of the response that is required under operational conditions, i.e. the "process" demand and subsequent plant response. This chapter is designed to give the reader some historical background and insight to the properties and applications of the new generation of magnetoelastic materials, in particular magnetostrictives.

The transduction property of either magnetostrictive or piezoelectric materials is the means by which the energy in magnetic induction or electric polarisation is converted into mechanical movement. Both types of materials form the backbone of active transducer technology. The transductive properties of magnetostrictive materials have significantly improved since the early days since the discovery and understanding of the magnetic and magnetoelastic properties of the rare-earth elements. Alongside the advances in permanent magnet technology, novel bulk magnetostrictive intermetallics containing iron, terbium and dysprosium have become available in commercial quantities. The potential of these improved magnetoelastic properties are now being harnessed in a wide variety of applications, such as sonar, valve actuation, and active vibration control [1; 2].

One of the best of the new magnetostrictive materials is known as Terfenol-D®, and typically $Dy_{0.7}Tb_{0.3}Fe_{1.95}$, generates strains in the region $\sim$ 1400ppm. The evolution of the compound originated from work on the elementary rare earths such as terbium (Tb), which at low temperatures ($< 200K$) generates huge strains in the order of 1%. Alloying with iron (Fe) enables room temperature operation, the strain capability can be realised in fields that are practically viable by the incorporation of dysprosium (Dy) [3] to compensate, reducing the magnetocrystalline anisotropy.

Other forms of these materials have also been investigated, i.e. amorphous ribbons and wires and thin films, as for many years there has been considerable interest in producing lower dimensional materials in order to reduce the size and/or the enhance specifications of devices for passive or active applications.

The impact of stress on these materials is of a fundamental nature both in terms of understanding the origins of magnetoelastic effects but, in some sense more importantly, on how these materials react operationally.

7.1.1 *Background*

The magnetostrictive effect, or Joule effect [4], refers to the change in physical dimension of a magnetic material as it is magnetized or its magnetic state is changed (volume conserving). In general magnetostriction is divided into two basic forms, linear magnetostriction and volume magnetostriction. As named, the former describes the change in length upon a change in magnetic state and the latter a change of volume. As volume magnetostriction is always much smaller than linear magnetostriction, the term "magnetostriction" generally refers to 'linear' magnetostriction (λ).

There is often confusion also between the terms magnetostriction and magnetoelasticity, the distinction being that the latter can be observed as a differing variety of responses not only as 'magnetostriction', but other effects such as changes in elastic moduli with the application of a magnetic field, the DE effect. All such effects originate from the interplay between the elastic and the magnetic energies of a material.

Magnetostriction has its formal roots in the fact that the anisotropy of a magnetic material varies with strain, i.e. changes in inter-atomic distance, and correspondingly the magnetic material lattice (ferromagnetic) will distort/strain spontaneously if the resultant strain lowers the ansiotropy energy. Thus any magnetic anisotropy theory based in terms of inter-atomic interactions will, in principle, be applicable as a theory on magnetostriction as well (Kittel [5]). Distinction also has to be made between the basic magnetoelastic coupling, as described above, and the more experimentally derived magnetomechanical coupling which denotes the efficiency conversion between magnetic and elastic energy densities, i.e. $\frac{\mu H^2}{2}$ to $\frac{E\epsilon^2}{2}$ respectively, and *vice versa* (μ is the permeability, H is the magnetic field, E the Young's modulus, and ϵ is the elastic strain).

Magnetostriction was first observed by James Prescott Joule in 1842 [4] where he established that iron, upon magnetization, increases in length in the direction of a magnetising field and contracts at right angles accordingly. In 1882, Barret [6] showed that nickel contracts in applied fields and, for the first time, observed definite changes in volume. During the first half of the 20th century, samples of iron, cobalt, and nickel, and various alloys, were studied at high and low temperatures. Single crystals studies were also carried out for the first time during this period. However, interest waned in the studies of magnetostriction [7], possibly due to the small magnetostriction constants being found (around tens of ppm). Table 7.1 shows a comparison between some early polycrystalline and single crystal magnetostrictions.

In terms of materials for applications, nickel or nickel-based alloys were considered the best up to the last quarter of the last century due to their relatively high strains and low magnetic anisotropies. However, in the later part of the century, these materials were overshadowed by competition from piezoelectrics and electrostrictive compounds which could operate more effectively at higher frequencies (e.g. lead zirconate titanate (PZT), Table 7.1).

Other magnetostrictive materials were developed during this period, e.g. cobalt-based ceramic ferrites, with improved strains and other properties like higher ohmic resistivity, to try and combat the challenge from the

Table 7.1 Comparison between early single crystal, polycrystalline magnetostrictive materials and an example of an electrostrictive (λ_{100} and λ_{111} single crystal coefficients and λ_s polycrystalline coefficient) [11; 12].

	$\lambda_{100}(\times 10^6)$	$\lambda_{111}(\times 10^6)$	$\lambda_s(\times 10^6)$
Single Crystal			
Iron (Fe)	21	−21	-
Nickel (Ni)	−46	−24	-
Polycrystalline			
Nickel (Heat Treated)	-	-	300
45 Permalloy	-	-	120
2V Permendur	-	-	1540
Electrostrictive			
PZT III	-	-	< 1000

dielectrics, but problems arose due to large magnetocrystalline anisotropies and low mechanical strengths.

The situation changed sharply in the early nineteen sixties due to the discovery of very large magnetic moments and other magnetic phenomena, including magnetostriction, in the heavy rare-earth elements, i.e. Gd, Tb, Dy, Ho, Er, etc. [8; 9; 10]. Problems with regard to applications still existed, however, as most of the rare earths display their high spontaneous magnetization at very low temperatures ($< 50K$), showing only negligible paramagnetic behaviour at room temperature, far above their Curie point. They also display very high magnetocrystalline anisotropy energies and hence require very strong applied magnetic fields to change their magnetisation distribution. Due to the combined effects of the above, it was very difficult to take advantage of the inherent large magnetostrictions for technical applications. Solutions based on intermetallic compounds or alloys of rare earth and other metals as R–Zn, R–Al and R–Fe series attracted attention and were widely investigated.

Magnetisation of a material is affected by external influences such as temperature and stress, compressive (−ve) or tensile (+ve). If a uniaxial compressive stress is applied to a sample, the magnetisation tends to align perpendicular to the direction of the applied stress if the magnetostriction is positive. If a tensile stress were applied on a positive magnetostrictive material then the magnetisation would tend to align parallel to the stress axis. For negative magnetostrictive materials the converse would occur. Since the application of stress changes the 'easy' and 'hard' directions, an additional term to the magnetocrystalline anisotropy energy is created. This is the magnetoelastic energy and in its simplest form, i.e. for a polycrystalline material is given by:

$$E_\sigma = -\frac{3}{2}\lambda_S \sigma \cos^2 \phi, \qquad (7.1)$$

where λ_S is the "saturation" magnetostriction, σ the applied stress and ϕ is the angle between the stress and the 'easy' direction of magnetisation. For the magnetostrictive materials considered here, σ is compressive and the magnetostriction positive. So that if the magnetisation is measured along the stress axis the sample would become harder to magnetise for increasing stress. The effect of stress becomes more acute when the magnetostriction is large.

The magnetomechanical coupling coefficient (k) is a useful transducer parameter as it may be an indicator of the efficiency of the transducer material. A magnetoelastic material, when subjected to a change in magnetic state, undergoes a change in mechanical (or elastic) state or *vice versa*. This coefficient is defined as the ratio of efficiency of conversion of magnetic energy to mechanical/elastic energy or *vice versa*. From equations of state assuming small signal regime, it can be shown that:

$$k^2 = \frac{d^2}{\mu^\sigma S_H}, \qquad (7.2)$$

where d is the piezomagnetic strain coefficient, μ^σ is the permeability at constant stress and S_H is the elastic compliance at constant field. Normally, k is obtained using a resonance technique, calculated from measured resonance and anti-resonance frequencies [13]. This analysis, however, has limitations due to the fact, that it does not consider all losses such as mechanical losses at resonance due to internal friction within the sample. Prediction of the losses in magnetostrictive materials and devices has been studied using plane wave modelling (PWM) [14].

7.2 Rare Earth Intermetallics

In 1963 (and following years) the basal plane magnetostrictions of the hexagonal rare earth elements terbium (Tb) and dysprosium (Dy) were measured at low temperatures and were found to be of the order 100 to 100,000 times typical transition metal magnetostrictions. However, due to their low Curie temperatures, the utilisation of these elements to room temperature applications could not be realized. A search was initiated in the 1970s to develop materials with giant magnetostrictions at room temperature. To this end the rare earth elements (R) were combined with known room temperature magnetic elements or compounds.

Among the R–Zn alloy series, DyZn single crystals exhibit a very large magnetostriction along their easy magnetisation direction and are good candidates for actuator applications at low temperatures ($< 77K$). Out of the range of rare earth-based intermetallic compounds, the R–Fe$_2$ system exhibits the strongest exchange interactions and highest ordering temperatures (far above room temperature) associated with the high iron content. The room temperature magnetostrictions displayed by such compounds are the largest ones ever found in any magnetic material, but they also display large inherent magnetocrystalline anisotropies. Magnetic properties of R–Fe$_2$ alloys are showed in Table 7.2.

Table 7.2 shows that TbFe$_2$ exhibits the highest room temperature magnetostriction, but has a very large magnetocrystalline anisotropy energy, with $k_4 = -7.6 \times 10^6 \mathrm{Jm}^{-3}$ (k_4 is a measure of the total anisotropy energy), and as such needs higher than a $10\mathrm{MAm}^{-1}$ applied magnetic field in order to reach saturation. On the other hand, the magnetic behaviour of DyFe$_2$ and Ho-Fe$_2$ compounds differ noticeably from TbFe$_2$ in that the direction of easy magnetisation is the [001] axis, i.e. the magnetocrystalline anisotropy energy is positive. Although it is three times smaller than that of TbFe$_2$,

Table 7.2 Polycrystalline magnetostrictions (λ_s at an applied field of $2\mathrm{MAm}^{-1}$), easy directions and a measure of the total anisotropy energy (k_4) found in some rare earth-iron$_2$ compounds [7; 15].

Rare Earth Iron$_2$	Easy Axis	λ_s at room temp. ($\times 10^6$)	k_4 ($\times 10^6 \mathrm{Jm}^{-3}$)
SmFe$_2$	[011]	-1560	-
TbFe$_2$	[111]	1753	-7.6
DyFe$_2$	[001]	433	2.1
HoFe$_2$	[001]	80	0.6
ErFe$_2$	[111]	-300	-0.33
TmFe$_2$	[111]	-125	-0.05

the anisotropy energy of $DyFe_2$ remains very large ($k_4 = 2.1 \times 10^6 Jm^{-3}$) compared to typical transition metal systems.

At first thought, small anisotropy energies and a large magnetostriction would seem incompatible. However, in the research studies undertaken by Clark [3], it was found that by alloying two rare earth metals of opposite sign of magnetic anisotropy energy, but of the same sign of magnetostriction, into a pseudo-binary compound, the anisotropy energy can be markedly reduced. They recognised that Tb and Dy make the perfect rare earth element pair, and they succeeded in preparing the compound $Tb_{0.27}Dy_{0.73}Fe_2$, a kind of ternary alloy. This alloy came to be known as Terfenol-D, *Ter* for terbium, *fe* for iron, *D* for dysprosium, and *nol* standing for the Naval Ordnance Laboratory, Washington DC, where the breakthrough was made.

As stated the choice of any smart material depends on its response to operational parameters, hence for magnetostrictives, this will be their magnetoelastic response to applied fields, mechanical loading and temperature.

Terfenol-D is available in grain-oriented form as commercial quantities of single crystals are not yet available. By the appropriate zoning of the Tb–Dy–Fe start alloy, grain growth along the $[11\bar{2}]$ crystal axis is encouraged. With the material having $\langle 111 \rangle$ easy axes and a positive magnetostriction, the result of grain alignment in the material is that the magnetostrictive response now depends not only on the strength of the applied field but also on any uniaxial stress [16] applied along the grain growth axis.

Figure 7.1 shows the field dependence of Terfenol alloy under varying applied compressive stresses. When magnetised by an external magnetic field, the material is further strained in the direction of the field ($\lambda_{\text{parallel}}$). On the other hand, when submitted to an external stress, the λ value becomes a function of the pre-stress. As the magnetostriction is positive for Terfenol-D, a compressive stress will cause an increase and a "sharpening" of developed strain. It also indicates how a low applied compressive stress may modify the strain–field (λ–H) curve. Hence from a practical point of view, actuation can be obtained efficiently with minimisation of the losses from the coil. However, a high compressive stress will reduce the magnetostrictivity ($d_{33} = \frac{\partial \lambda}{\partial H}$) (Fig. 7.2). In addition, the magnetoelastic properties may vary markedly depending on the operating conditions and material manufacturing techniques [17; 18; 19].

The application of these material is not as straightforward as it appears as the strain–field curves are nonlinear and hysteretic (Fig. 7.1), i.e. lossy.

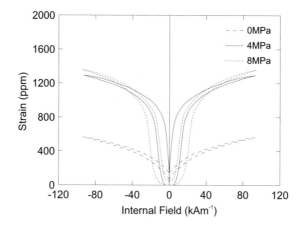

Fig. 7.1 Typical magnetostrictive strains for Terfenol-D as a function of internal magnetic field (internal field — applied field corrected for demagnetisation effects) for varying applied pre-loads.

To obtain optimum transduction performance both field and uniaxial pre-stress need to be adjusted. Hence, optimum magnetomechanical coupling (k) or d coefficient (d_{33}) is obtained under specific field and pressure conditions (Fig. 7.2). The magnetomechanical coupling, measured on a variety of samples, is usually in the region of 0.7, i.e. giving transduction efficiency of $\sim 50\%$. The use of PWM has indicated that the experimentally derived values of k are as much as 15% higher than those calculated from resonance even with eddy current correction. This is attributed to frequency dependent losses, which not only influence the magnitude of the resonance and anti-resonance frequencies but the shape of the whole resonance curve.

For the present materials, a uniaxial compressive strength $\sim 350\text{MPa}$ is about the upper limit but this has been doubled with the addition of aluminium to the rare-earth-iron alloy as shown in Table 7.3 [20]. A magnetostrictive device, due to its inductive nature is ideally suited to low frequency operation, even down to DC. Eddy current effects (loss) determine the upper frequency limit, but Terfenol-D is available in laminated form and in theory is capable of operation up to about 80kHz. Some assisted cooling may be required due to eddy currents under very high drive conditions, but modifications to the composition have raised the electrical resistivity (Table. 7.3).

Table 7.3 Comparison of parent Terfenol-D and aluminium substituted material.

	$Tb_{0.3}Dy_{0.7}Fe_{1.95}$	$Tb_{0.3}Dy_{0.7}(Fe_{0.9}Al_{0.1})_{1.95}$
λ_{max} at 120kAm^{-1} (ppm)	1380	1160
k_{max}	0.7	0.63
Resistivity ($\mu\Omega$cm)	63.1	97.2
Compressional strength (MPa)	320	735

The effect of temperature on the anisotropy is also important since thermal energy affects the magnetoelastic response of these compounds. Estimates have been made of the change in stress-induced anisotropy and the magnetocrystalline anisotropy between parent and aluminium substituted compounds [20]. Figure 7.3 shows temperature dependence of magnetostriction of $Tb_{0.3}Dy_{0.7}Fe_2$ under a compressive stress of 18.9MPa [21]. The temperature dependence of magnetostriction near 20°C is $\frac{\partial \lambda}{\partial H} = 5.6°C^{-1}$. The operational range is composition dependent but is typically $-10°C$ to $+120°C$ or higher with small additions of cerium [22].

The elastic or other mechanical properties are important for the application of Terfenol-D. Figure 7.4 shows the field dependence of Young's modulus of Terfenol-D [23]. Because magnetostriction strain is associated

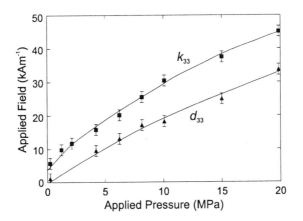

Fig. 7.2 Magnetic field and pressure magnitudes required for either the optimum magnetomechanical coupling (k_{33}) or optimum magnetostrictive strain coefficient (d_{33}) for a sample of Terfenol-D [12].

with the rotation of magnetic moments caused by external stress, the strain created by external stress differs from the one predicted by Hooke's law, ΔE effect [24]. So under the application of large magnetic fields, the Young's modulus reaches an upper limit. Figure 7.5 gives the relation between displacement and output force for two different actuators with the same dimensions: one of Terfenol-D and one of laminated PZT [25]. Both displacement and maximum output force of Terfenol-D are markedly increased in comparison to the laminated PZT. These actuators deliver high displacements, with a high energy density and strong force capability, in a broad frequency bandwidth. A load of 180kg may be displaced with a stroke as large as 100μm using commercially available actuators.

In 1996 and 1997, experiments carried out by Prajapati et al. [25; 26] at the University of Hull showed that, under a cyclic uniaxial stress of ±1000MPa about a mean pre-load of 120MPa along the samples growth axis, the magnetostrictive strain and d_{33} coefficient for grain-oriented Terfenol-D increased by approximately 56% after the initial 10^6 cycles of stress and reached a maximum of $\sim$ 1660ppm after 28.5×10^6 cycles. The induced stress anisotropy led to a peak in the field dependence of the permeability coincident with the enhancement of the magnetostrictive strain coefficient d_{33}. These "surprisingly good results" are markedly very different

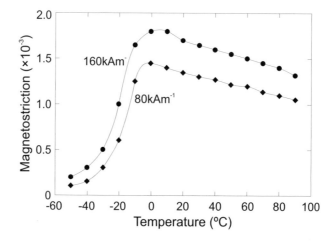

Fig. 7.3   Effect of temperature on the magnetostriction generated by Terfenol-D at two constant values of applied magnetic field [20].

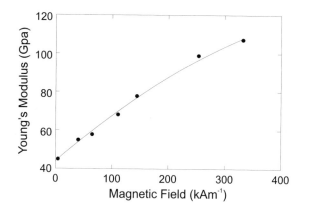

Fig. 7.4   Variation in Young's modulus for Terfenol-D with applied magnetic field [23].

from the behaviour of piezoelectrics and electrostrictives such as PZT where breakdown under such cyclic operation can occur.

It had been reported that the stress cycling will cause degradation effects in the conductive and mechanical properties. The above results challenged the tradition consideration that stress cycling imposed on Terfenol-D during the course of its working life in an actuator might lead to a deterioration in

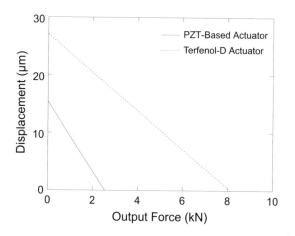

Fig. 7.5   Displacement as a function of output force for a PZT based actuator and a Terfenol-D actuator of similar dimensions (100mm^2 × 18mm) [24].

its magnetostrictive and magnetomechanical properties and eventually to complete mechanical failure. So it shows that magnetostrictive devices constructed from Terfenol-D can offer a significant advantage compared with ceramic PZT actuators when applied in load bearing situation. Prajapati considered that the ultimate origin of the effects lies with the exceptionally high magnetoelastic coupling in Re–Fe$_2$ compounds as Terfenol-D, but will only occur in single crystal samples or polycrystalline materials with sufficient grain orientation.

7.3   Actuation

Terfenol-D is considered to be more powerful than similar types of smart materials due to its high strain and high force capability, wide bandwidth and microsecond response times. At present magnetostrictive materials account for around 10% of the active applications market. The transduction properties of magnetostrictive materials have significantly improved since the early days originating from the discovery and understanding of the magnetic properties of the rare-earth elements. The potential of these improved magnetoelastic properties are now being harnessed in a wide variety of applications.

7.3.1   *Generic Actuators*

Controlled actuation can be achieved if the performance of a Terfenol-D device and its application are matched. However, as previously indicated difficulties can arise from hysteresis and reaction to static and dynamic loads. The premise is to design a device to obtain the maximum strain from the active element with the minimum electrical power input. A static magnetic field is required to bias the material at a mean operating point. The choice of the biasing point allows the device to operate in the 'linear' region of its strain–field curve and to be efficient by harnessing greatest d_{33}. This biasing field is usually supplied by permanent magnets, thus minimising the need for electrical bias current and so reducing power requirements. The driving magnetic field is supplied by a drive coil of sufficient dimensions wound in close proximity to the active element. The active material must also be pre-stressed so as to bias it on the optimum magnetostriction curve for each given application (Fig. 7.6).

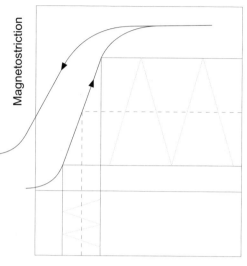

Fig. 7.6 Application of driving AC magnetic field around DC biasing point for the generation of actuator displacement.

Actuators generally always consist of several components; an active element, i.e. Terfenol-D rod, a pre-stressing mechanism e.g. a spring, a permanent magnet or DC electric coil for static magnetic field biasing, drive field coil, cooling system (for some kind of designs), support structure and output connection (Fig. 7.7). Other actuator layouts have been developed such as a "push–pull" actuator by using a pair of opposing Terfenol-D rods working in a symmetrical way [27]. Actuator costs are always a major consideration, but with the general high price of Terfenol-D it is becomes even more so. Although work is being carried out to try and to simplify the manufacturing process for Terfenol-D, the cost of rare earth metals may be difficult to reduce.

Following these premises even an actuator has been designed and built to enhance the drilling of rocks such as sandstone, limestone or granite capable of delivering at one extreme several hundred microns of movement, and, at the other, forces in excess of 12kN (Figs. 7.8 and 7.9) [29].

7.3.2 *Magnetostrictive Motors*

Many attempts have been made to construct motors based on magnetostrictives trying to utilise the active materials' very compact, precise and powerful qualities. There are many kinds of designs and some have already been working at the research level. The advantages of these kinds of motors include: high force density, high stiffness and high precision. Compared with most electromagnetic servomotors, these motors can overcome two major disadvantages of former: backlash resulting from gear mechanisms and increased torque; and non-self-locking with cut-off of zero power supply.

The first kind of motor is the Linear Motor or Kiesewetter motor [30]. This usually consisting of three parts: (1) a stator which is a stiff cylindrical tube made of stainless steel; (2) a Terfenol-D shaft which is inserted into the stator under a magnetic field; and (3) a number of short magnetising coils which may create a magnetic field on a small part of the rod (Fig. 7.10). Here the linear sequencing of the coils in turn alternately "unclamps" and "clamps" sections of the active element creating an "inch-worm" effect. These motors are capability of high force, larger displacements than normal and high precision. This kind of motor can be used in such application areas as manufacturing, force sensing, aviation control devices and process control.

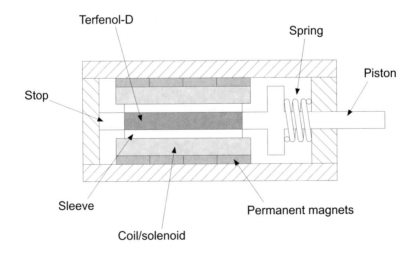

Fig. 7.7 Generic schematic of magnetostrictive actuator layout and components [28].

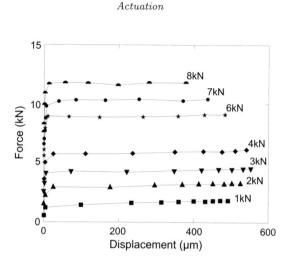

Fig. 7.8   Displacements and forces developed under a compliant load (spring constant $k = 3 \times 10^5 \text{Nm}^{-1}$) [29].

A second kind of motor is the so-called Hybrid Terfenol–PZT motor. It uses a shear-free clamping mechanism by using laminated PZT and an active Terfenol-D actuator element [31]. It has a higher frequency response up to 300Hz.

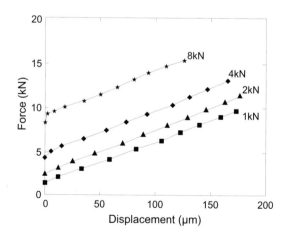

Fig. 7.9   Forces developed against a compliant load ($k = 3 \times 10^7 \text{Nm}^{-1}$) as a function of displacement for different levels of pre-stress [29].

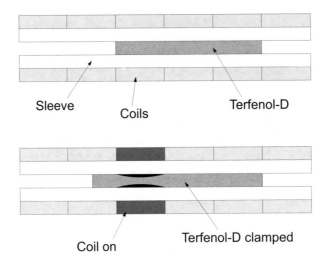

Fig. 7.10 Schematic layout and principle of operation of a basic Kiesewetter or elastic wave motor.

A third example of magnetostrictive servomotors is the Direct Drive Rotary motor. It is designed to satisfy the requirement from the aerospace industry looking for new motors with high energy density and higher torque at low speeds [12; 32].

7.3.3 Sonic and Ultrasonic Emission

Mainly due to military requirements the application of Terfenol-D for sonic and ultrasonic emission is becoming more and more important and is well developed in many countries, its main usage being for sonar. Until now, most sonar systems use Tonpilz-type piezoelectric transducers. At high frequencies this type of actuator can satisfy most requirements very well. However, to increase the sonar ring, it is better to work at lower frequencies and the size of the piezoelectric transducer then becomes a major problem. Due to it's large magnetostriction, larger energy density and lower Young's modulus, Terfenol-D is deemed the best replacement active element. For the same configuration and volume as a piezoelectric transducer, the emitted acoustic energy of a Terfenol-D transducer shows a noticeable increase. Typically all these actuators operate in the resonant condition [33; 34].

7.3.4 *Vibration Control and Absorbers*

Active vibration control is one application which can utilise magnetostrictive actuators. Here, the ability to control actuators in spite of their nonlinear behaviour and hysteretic nature is essential.

In 1999, Pratt *et al.* [35] introduced a Terfenol-D nonlinear vibration absorber. It showed that non-biased Terfenol-D actuators can perform effectively a "squaring" operation, instead of the applied algorithm, and achieve similar results. The results indicate that the saturation-based control technique implemented with a Terfenol-D actuator constitutes an effective nonlinear vibration absorber.

Considering the advantages of Terfenol-D as simplicity and reliability, low mass, low voltage, and insensitivity to centripetal acceleration, Fenn *et al.* [36] developed Terfenol-D driven flaps for a helicopter vibration reduction actuator for a high-weight-penalty helicopter application. The magnetostrictive actuation system weighs less than 1% of gross vehicle weight, and use only 0.7% of cruise power. The other required subsystems of the vibration reduction system are available from commercial sources or are described in the literature. Helicopter vibration reduction greater than 90% is predicted because of the superior actuator performance and resulting individual blade control.

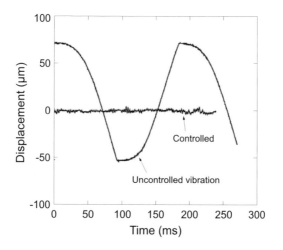

Fig. 7.11 Vibration levels at 5Hz before and after the implementation of DVSC active vibration cancellation [38; 39].

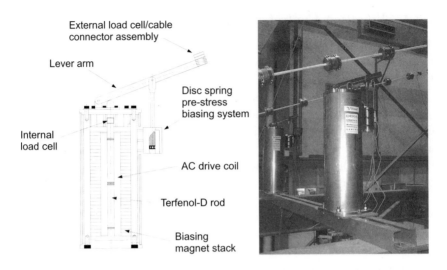

Fig. 7.12   Schematic and photograph of high force prototype magnetostrictive actuators used in the cable-stayed bridge mock vibration damping research part of the Brite–Euram ACE project [40].

An adaptive vibration control system using a Terfenol-D actuator with was demonstrated by Shaw [37] in 1998. It was used to provide a negating force to an external excitation for suppressing vibration response of the disturbed system. Experimental results indicate that over 96.6% (or 29.3dB) of the vibration displacement amplitude of the disturbed system is attenuated in a tested frequency range of 50–70Hz, thus demonstrating the validity and effectiveness of the plant disturbance cancellation.

The conjunction of Digital Variable Structure control (DVSC) techniques with Terfenol-D actuators not only contends with the characteristics of the actuator, which can differ slightly from one to another, but also to adapt to changes in actuator response arising from static and dynamic loads. To date, 32dB of cancellation at 5Hz has been achieved [38; 39] (Fig. 7.11).

Prototype large scale, high force actuators have been designed and manufactured for research into the control of cable vibrations in a large-scale cable-stayed bridge mock-up [40]. Studies have verified the capability to provide structural damping by means of an actuator–sensor feedback control system. By exciting the bridge mock-up system at resonant frequencies, large amplitude cable vibrations (46mm, 0.8kN peak-to-peak) are produced.

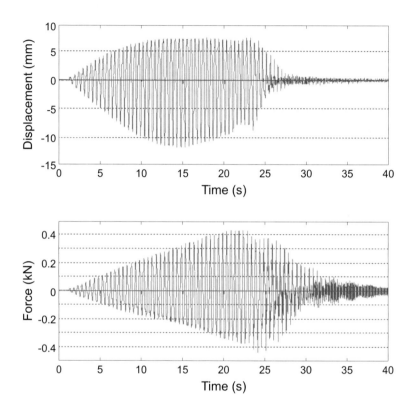

Fig. 7.13 Displacement and force transducers output for controlled vibrations in bridge mock-up (control switched on after 25 seconds) [40].

These can then be successfully mitigated by use of the active control actuators with an increase of the structural damping factor by an order of magnitude, using minimal power requirements, Figs. 7.12 and 7.13. The research has also highlighted the potential of utilising such actuators as diagnostic health monitoring tools in civil engineering structures.

7.4 Conclusions

Future magnetostrictive-based actuation technology success depends on a few critical factors. In addition to successful production of the actuators, drive amplifiers and controllers that are already available, these have to be engineered to meet particular applications according to space, displacement, force, speed and power specifications. Knowledge of how the materials react to external influences is crucial to achieving the above.

Acknowledgements

The authors are grateful to colleagues at the Universities of Salford and Hull for their support and encouragement in the preparation of this chapter and to Newlands Technology Ltd. (Newlands Scientific PLC). Financial support from the EPSRC and EU programmes is also gratefully acknowledged.

Bibliography

[1]   Greenough, R. D., Jenner, A. G., Schulze, M. and Wilkinson, A. J. (1991) *J. Magn. & Magn. Mat.*, **101**, 75–80.

[2]   Greenough, R. D., Jenner, A. G., Schulze, M. and Wilkinson, A. J. (1991) *IEEE Trans. on Magnetics*, **27**, 5346–5348.

[3]   Clark, A. E. (1974) *AIP Conf. Proc.*, **18**, 1015.

[4]   Joule, J. P. (1842) *Phil. Mag.*, 930, 30, 76 and 225.

[5]   Kittel, C. (1949) *Rev. Mod. Phys.*, **21**, 555.

[6]   Barret, W. F. *Nature*, **26**, 585–586.

[7]   Lacheisserie, E. de T. (1993) *Magnetostriction — Theory and Applications of Magnetoelasticity*, CRC Press.

[8]   Corner, W. D. *et al.* (1960) *Proc. Phys. Soc.*, **75**, 781.

[9]   du Plessis *et al.* (1968) *Phil. Mag.*, **18**, 145.

[10] Rhyn, M. (1965) *Phys. Rev. A*, **140**, 2143.

[11] Greenough, R. D. and Schulze, M. P. (1994) *Intermetallic Compounds* (ed. J. H. WestBrook and R. L. Fleisher), vol. 2, Wiley.

[12] Jenner, A. G., Smith, R. J. E. and Greenough, R. D. (2000) *Mechatronics*, **10**, 457–466.

[13] Savage, H. T., Clark, A. E. and Powers. J. (1975) *IEEE Trans. on Magnetics*, **11**, 1355.

[14] Reed, I. M. (1994) PhD Thesis, University of Hull, UK.

[15] Clark, A. E. (1980) *Ferromagnetic Materials* (ed. E. P. Wohlfarth), vol. 1, North Holland.

[16] Clark, A. E., Teter, J. P., and McMasters, O. D. (1988) *J. Appl. Phys.*, **63**, 3910.

[17] Verhoven, J. D., Gibson, E. D., McMasters, O. D. and Baker, H. H. (1987) *Metall. Trans. A*, **18**, 223.

[18] Galloway, N., Greenough, R. D., Schulze, M. P. and Jenner, A. G. (1993) *J. Magn. & Magn. Mat.*, **119**, 107.

[19] Sandlund, L. and Cedell, T. (1992) *Proc. 3rd Int. Technology Transfer*

Congress, Actuator 92, Bremen, Germany.

[20]   Jenner, A. G., Prajapati, K. and Greenough, R. D. (1993) *IEEE Trans. on Magnetics*, **29**, 3514.

[21]   Clark, A. E. (1992) *Proc. 3rd Int. Conf. on New Actuators, Berlin, Germany*, pp. 127.

[22]   Patent Pending, Jenner, A. G. *et al.*, University of Hull, Hull, UK.

[23]   Clark, A. E. and Savage, H. T. (1975) *IEEE Trans. on Sonics and Ultrasonics*, **22**, 50.

[24]   Akuta, T. (1989) *Proc. 10th Int. Workshop on Rare Earth Magnets and their Applications, Kyoto, Japan*.

[25]   Prajapati, K., Greenough, R. D. *et al.* (1996) *IEEE Trans. on Magnetics*, **32**, 4761–4763.

[26]   Prajapati, K., Greenough, R. D. *et al.* (1997) *J. Appl. Phys.*, **81**, 15.

[27]   Kvarnsjo, H. *et al.* (1992) *Proc. 3rd Int. Conf. on New Actuators, Berlin, Germany*.

[28]   Halkyard, P. and Holiday, P. (2001) Private Communication, Newlands Technology Ltd.

[29]   Aston, M., Greenough, R. D., Jenner, A. G., Metheringham, W. J. and Prajapati, K. (1997) *J. Alloys & Compounds*, **258**, 97–100.

[30]   Int. Patent W088/05618

[31]   Butler, J. L., Butler, A. L., Butler. S. C. (1993) *J. Acoust. Soc. Am.*, **94**, 636–641.

[32]   Claeyssen, F., Lhermet, N., LeLetty, R. and Bouchilloux, P. (1997) *J. Alloys & Compounds*, **258**, 61–73.

[33]   Moffet, M. B., Clarke, A. E. *et al.* (1991) *J. Acoust. Soc. Am.*, **89**, 1448.

[34]   Moffet, M. B., Powers, J. M. and Clake, A. E. (1991) *J. Acoust. Soc. Am.*, **90**, 1184.

[35]   Pratt, J. R., Oueini, S. S. and Nayfeh, A. H. (1999) *J. Intel. Mat. Sys. & Struct.*, **10**, 29–35.

[36]   Fenn, R. C., Doener, D. A. *et al.* (1996) *Smart Mat. & Struct.*, **5**, 49–57.

[37]   Shaw, J. (1998) *J. Intel. Mat. Sys. & Struct.*, **9**, 87–94.

[38]   Jenner, A. G., Greenough, R. D., Allwood, D. and Wilkinson, A. J. (1994) *J. Appl. Phys.*, **76**, 7160.

[39]   Smith, R. J. E., Jenner, A. G., Wilkinson, A. J. and Greenough, R. D. (1997) *J. Alloys & Compounds*, **258**, 101–106.

[40]   ACE (Active Control in Civil Engineering) Project funded by the European Community under the Industrial & Materials Technologies Programme Brite Euram 3, proposal N.BE96-3334, contract BRPR-CT97-0402 (1998).

[41]   Wickenden, D. K., Kistenmacher, T. J., Osiander, R. *et al.* (1997) *Johns Hopkins APL Technical Digest*, **18**, 271–278.

[42]   Dooley, J. and DeGraef, M. (1997) *Ultramicroscopy*, **67**, 113–131.

[43]   Geng, Z. J., Pan, G. G., Haynes, L. S., Wada, B. K. and Garba J. A. (1995) *J. Intel. Mat. Sys. & Struct.*, **6**, 787–800.

[44]   Hardee, H. C. and Hardee, N. L. (1997) *J. Alloys & Compounds*, **258**, 83–86.

Chapter 8

Smart Fluid Machines

William A. Bullough

Department of Mechanical Engineering,
University of Sheffield,
Mappin Street, Sheffield S1 3JD, UK.

8.1 Introduction

The aim of this chapter is to introduce the topic of smart fluid machines in layman's language. Successively, the objectives are to explain the concept of flexible machine operation, digital catching, load requirements on smart fluid power transmissions and the scope for smart fluid devices. Following two easy to understand kinematic applications the wider electronics–hydraulics interfacial field and commercial applications are described in brief.

8.2 Concepts and Philosophy

The smart machine features variable configuration of function (flexibility) via electronic control alone: the same hardware is always involved with no downtime for changing the geometry of the solid parts and hence the motion — via alteration of levers, gear changes, swapping pulleys, etc.

Large forces F, high speeds U and significant displacements δ are the prime features of a machine. In perspective, mechanical machines are the fastest and most forceful of all; they are inflexible yet are capable of high

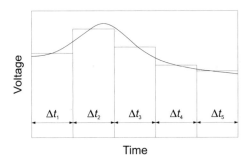

Fig. 8.1 'Busbar' catching analogy for flexible drive methodology.

precision and stiffness of hold over lengthy sweeps or strokes ∼ 250mm. Electromagnetic machines inherently feature variable speed working, are weaker than mechanical machines but have recently become more capable of digital or analogue flexible operation via servo or stepper motors. Their competitive position has significantly improved since the development of new permanent magnet materials and solid state controllers which can rapidly force current (producing magnetic field) into their windings.

Machines of all types are limited by heating, stress/flux saturation and, for flexible operation, their in-built forcing stress/mass ratio determined speed of changing speed or, as we say, inertial loading and consequential distortion of the machine parts or load. Indeed there is literally no time for a flexible machine to stop and start in a conventional sense — the input or brake must already be available for 'catching on to'. This passive process can perhaps be best demonstrated by analogy with the electronic/voltage function generator, see Fig. 8.1.

Here 'busbars' of electrical tension are latched or 'catched' onto in turn, for differing time intervals Δt. Any attached load must be capable of fast response in both volts rising and falling directions. In the smart fluid machine the fluid provides the electronically controlled catching element. The use of a liquid additionally conveys with it the usual advantages of a hydraulic power transmission. However, before going into further philosophical detail with respect to "why do we need" a smart fluid catch, the demand or customer perspective is introduced via relatively slow motion examples.

The nematic liquid crystal based (70% by volume) electrorheological fluid employed in the adaptive walking frame (Fig. 8.2(a)) provides an uncomplicated example of the matching of fluid properties to flexible machine requirements via electronic control. When an excitation of $\sim 2\text{kVmm}^{-1}$ is applied between electrified plates separated $\sim 0.5\text{mm}$ by the mixture (as in the brake of Fig. 8.2(b)) the liquid crystal domains polarise and tend to align themselves according to the field direction with a polymer film (30%

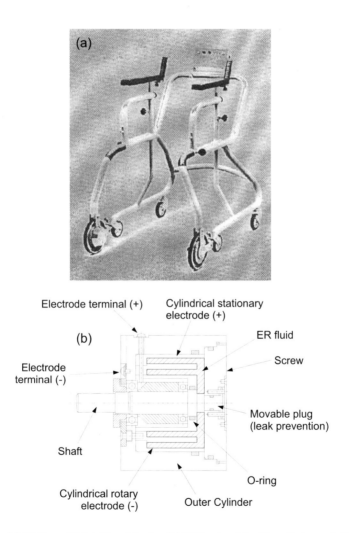

Fig. 8.2 (a) Walker with intelligent brake; (b) ER brake of double cylinder and electrode type.

by volume) building stiffening chains or bridges between the domains. Due to the size of the domains and the high inherent viscosity of the 'homogeneous' mixture $\sim$ 10Pas, the time to full rheological response from voltage application is relatively slow $\sim$ 20–80ms but the limiting electronically con-

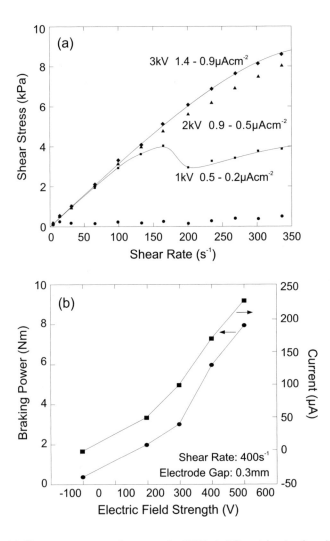

Fig. 8.3   (a) Shear stress versus shear rate for ERF at different levels of applied excitation; (b) characteristics of Asahi ER brake.

trolled stress level is high $\sim 10\text{kPa}$ at 0.3ms^{-1} walk speed and 300s^{-1} fluid shear rate — see Fig. 8.3(a) [2].

The function of the Asahi Kasei Corporation frame is to limit the speed of any walker who uses it, thus preventing stumbling whilst allowing its adoption to the various terrain inclinations encountered and the weight of the walker. It does this via pulse width modulation of the high voltage applied to the brake as indicated by the speed sensors attached to the wheels of the frame.

This action is enhanced by the low current density demand of the liquid crystal siloxane $\sim 1\mu\text{Acm}^{-2}$ of electrode surface and results in a brake torque versus voltage characteristic which is quite different from that of the fluid (Fig. 8.3(b)) for a fraction of a Watt control power.

The brake, which measures some $15 \times 100\text{mm}$ dia., will handle a 60kg person on a $15°$ slope and will allow a fraction of the resistance required there when on the level. It will also produce a rigid restraint as necessary, e.g. at bedside. The seemingly unduly slow time response of the side chain liquid crystalline polysiloxane mixture is adequate for the purpose and the brake thermal load can be dissipated from the fluid without any great effect on the characteristic [1], though high temperature can severely reduce the ER effect.

The brake/frame unit is manufactured by the Asahi Engineering Co. in Japan and has been used with great effect by Parkinson's disease patients. Future applications of the brake type are expected on wheelchair, limb rehabilitation and lift assist devices.

A second medical application, the intelligent knee orthoses is expected to join the walking frame on the market soon, Biedermann OT Vertrieb of Germany are carrying out field trials on a Lord Corporation (USA) fluid filled valve/piston/cylinder attached to their artificial leg motion controller (Fig. 8.4) [3].

The prosthesis (knee replacement) system uses several strain gauges in the shin region to determine axial force, and movement sensors in the knee pivot to measure angle, direction and rate of swing. Based on these data a microprocessor is able to assess what the person wants to do (walking rate, tackle inclinations, ramps, stairs, standing, etc.) and heuristically determines the input to the hydraulic damper/controller (magnetorheological suspension filled cylinder) in order to optimise gait for that mode. The response time of the system and algorithm is about 40ms, which is similar

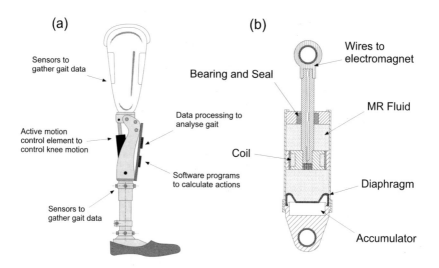

Fig. 8.4   (a) Above knee orthoses with MR damper; (b) schematic of MR damper.

to the response time for signals in the living knee. Battery life averages about 2 days on the magnetorheological fluid.

A third form of controller which can encompass large δ is the hydrodynamic bearing type.

Certain fundamental requirements on the fluid performance required, will have been noted from these examples: the ability of the leg to move freely implies a low effective viscosity μ_0 in one state (see Fig. 8.5(a)) whilst the necessity to resist stress after the fashion of a plastic is a property that may be called quickly — a plastic will not flow below an applied stress τ_e (which is variable dependant on the excitation level applied to the fluid in a cylinder by-pass or parallel flow controller, e.g. Fig. 8.5(b), above which flow of the electrostructured fluid (ESF) will again take place. Such "non-Newtonian" behaviour can occur only in mixtures of which one phase is a dispersed solid of significant ($> 1\mu m$) size in a liquid. A usable magnitude of yield stress depends on a relatively large concentration of solid by volume (20–60%). Its speed of response to an electric or magnetic polarising excitation and consequent realignment of the general particle mesh structure and velocity profile under the action of hydrodynamic and electrical forces will be enhanced by having particles of small size. This brings with it an ability to interface well with IT architecture — provided the general conductance of the fluid is low. It will be noted that the characteristics shown are stable in so much as if an overload causes the shear rate $\dot{\gamma}$ to increase then $\tau_e + \mu_p \dot{\gamma}$ increases in resistance to it.

A further convenient aspect of the electrostructured fluid is the ease by which a hydraulic type force transmitter can be built to achieve linear motion. This contrasts well with any mechanical device and the short-range nature of purely electrically activated strong linear force devices (solenoids, etc.). A judicious choice of the controller size determines [4] the electrical characteristics, and base fluid properties will provide lubrication at say a piston/cylinder interface whilst the relatively incompressible nature of a liquid will bestow high positional stiffness against a disturbing force applied to the piston rod, i.e. when the load produces a shear stress within the yield stress level dictated by the excitation field.

The production of a combination of rapid motion and large force is enhanced by the fact that hydraulic oils have a very high density × specific heat product: $\rho \times c_v \times \Delta T$ absorbs the frictional heat generated and inhibits rapid temperature excursions. In this respect the best fluids in order of thermal capacitance are water, oil, platinum, gold, silver, etc. — how lucky we are that the two best liquids in this respect are so plentiful. Finally, considering pressure acts in all directions and that the load determines the level of pressure generated, severe stress concentrations rarely occur in a hydraulic system, a valuable property compared to gear wheels, levers, cams, etc., when inertial/acceleration terms and regular switching are conducive to enhanced crack growth rate and fatigue.

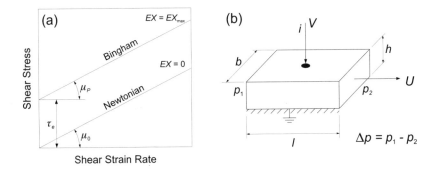

Fig. 8.5 (a) 'Ideal' MR/ER fluid characteristic at different excitation levels; (b) plane ERF control valve.

A simple force balance on the controller plates brings the shape and size of an arrangement into the equation:

$$\Delta pbh = 2\tau b. \tag{8.1}$$

Strictly steady state equations in turn show how a relatively small but controllable shear stress can result in the modulation of a sizeable piston (diameter d) force viz:

$$F = \Delta p \frac{\pi d^2}{4}, \tag{8.2}$$

or

$$F = 2\tau \frac{l}{h} \frac{\pi d^2}{4}. \tag{8.3}$$

For $h = 0.5$, $l = 100$, $d = 100$mm and $\tau_e = 100$kPa then

$$F \sim 4 \times 10^5 \text{N}.$$

In the highly unsteady state such limiting factors as the acceleration of the fluid in the control valve and compressibility of the fluid including any gas filled accumulator in the circuit can be detrimental to performance. Hydraulic type oil is some 150 times more compressible than steel and if the piston, with area A, were to accelerate at X then the acceleration in any connecting lines having area a would be $\frac{XA}{a}$.

From these examples there can be discerned the three requirements that are generally to be satisfied prior to the adoption of a smart fluid mechanism:

(1) Passive catching action with an input provided from a relatively high capacity source.
(2) A differing kind of adaptable outcome required, this to be determined by information technology and intelligent signal processing.
(3) The likelihood that a sophisticated and forceful result sought cannot be achieved economically or logistically by purely electrical or mechanical devices.

The satisfaction of these needs by other electrically controlled means is now explored.

8.3   More Philosophy

In an approach which is more general than that done via the above specific examples a 'periodic table' of machines seems to affirm the idea gleaned from Fig. 8.1, the digital electric catching concept.

Does the blank box indicate a digital electronic machine type? Throughout history the machine of the age invariably mimics its time piece [6]. The types of power transmission that differentiate the clock–machine pairs being respectively (in the different machine ages):

(1) Mechanical/hydraulic age: no variation without engineered geometry change, spring, and pendulum clock, etc.
(2) Electromagnetic age: speed variations without geometrical changes, synchronous motor drives electric clock.
(3) Information technology age: function change at the flick of a switch! Digital watch.

So far, the now expected paradigm in machines has involved the application of computers, sensors and IT operating strategies to machines of the first and second ages. Third age or smart technology attempts at new machine or power transmission creations via the liquid state devices presently being described recognise the heat removal and linear motion aspects of smart

Table 8.1   'Periodic table' of machine types.

Discipline	Static	Kinematic	Dynamic
Mechanical	Lever	Spinning machine, valve gear	Loom and forging press
Thermofluid	Combustion and heating	Water, heat, pneumatic transmission systems	Locomotive, servo controlled hydraulics
Electromagnetic	Transformer, magnetostrictor	Telegraphy, radar	Stepper motor, solenoid, etc.
Electronic	Piezoelectrics	Television, computer	?

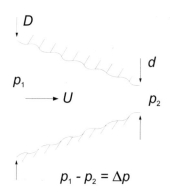

Fig. 8.6   Generalised fluid controller.

device provision. Solid state devices eventually cannot be cooled since the heat generated therein is proportional to the cube of semiconductor size whilst its surface area sizes only on square measure: the electrorheological fluid is a true hydraulic semiconductor — it reacts to electric fields and has a negative temperature coefficient of resistance. All ESFs can be moved out of the hot zone, cooled and recirculated, an important aspect of hydraulic machine design. Digital excitation is relatively inexpensive.

Pursuing the prospective linking of hydraulic flexibility, etc. and electronic passive catching by way of a semi-conductive method (of restricted current flow) to link in with the signal conditioning feedback, IT processing requirement and catching techniques, a dimensional analysis of a control element is undertaken:

$$\Delta p, \, \tau = f(\rho, \mu, U, d, D), \tag{8.4}$$

or

$$\frac{\Delta p}{\frac{1}{2}\rho U^2}, \, \frac{\tau}{\frac{1}{2}\rho U^2} = f\left(\frac{\rho U d}{\mu}, \frac{d}{D}\right). \tag{8.5}$$

Given that all hydraulic fluids possess roughly the same density ρ then the pressure and shear stress coefficients at any mean velocity of flow U may only be modulated by a change in geometry or viscosity μ, each of which must be electronically induced. The former is presently achieved by proportional servo-spool or solenoid valves respectively.

The servo-spool valve is expensive on account of the close tolerances on lands and ports diameters and lengths but can work well up to frequencies of 100Hz or so in terms of output flow into a light load per unit pilot motor input current. The solenoid is mainly an on/off device by virtue of the large coil/iron driver combination but produces a self cleaning and cheap poppet valve form. Together they demonstrate the problems of electro-hydraulics: the driver can be small and weak but fast, or strong and sluggish. Nevertheless the quest for fast hydraulic controllers is such that stepper motor driven screw down poppets or dithering solenoid drives are being applied.

8.4 The Strictor Driven-Hydraulic Valve

One solution of the problem which combines the best of the above characteristics with a solid state interface is to drive the poppet with some kind of electrically strictive device of the composition, e.g. of Fig. 8.7 — if the 'hydraulic busbar' is provided by a power pack or whatever [7].

At present the 0.1% of length extension of the many (out of scale in figure) glued together active elements of the piezo stack practically limits the throat opening to about 25% of the port area. However recent trends in piezo development and processing (less voids, better insulation and higher voltage operation) seems set to greatly increase this towards full opening for a $10d$ stack length (see Chapters 6 and 7 for composition, characteristics and *modus operandi* of piezo electric or magneto strictive driver elements).

8.5 Electrostructured Fluids

Returning now to the possibility of modulation of the viscosity of the hydraulic medium indicated in Fig. 8.6.

A dual flow-hold effect of the Bingham plastic type derives from the formation of particles into chains (Fig. 8.8) under the action of the applied field which causes the inter particle dipolar reaction. If the yield strain capacity set by the field is exceeded by the application of an external stress then flow will occur. If the particles are iron then a magnetising force H is needed to cause the polarisation. If the particles are semi-conductive polymers (or equivalent) or particles an electrostatic field is called for.

In the electrorheological fluid the means of polarisation is considered to be by surface ion or electronic transport set in motion by an electric field

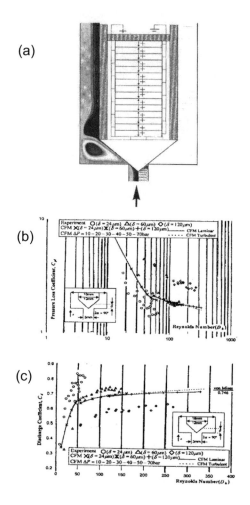

Fig. 8.7 (a) Piezo poppet valve CFD (out flow type) [after A. P. Wong]; (b) pressure loss coefficient versus Reynold's number based on hydraulic diameter for cone angle $2\alpha = 90°$; (c) discharge coefficient versus Reynold's number for $2\alpha = 90°$.

gradient of several thousand volts per millimeter usually applied between parallel plate electrodes. The solid can have a semi-conductive nature or be porous and carry water. The mismatch in dielectric properties between particle and dispersant is important to the structure inducing forces as is the amount of current trickle through the mixture. The magnetorheological suspension (MRS) will carry carbonyl–iron particles in a wide range of dispersants. The base of the mixture is not so important chemically as in an ERF — it can be from water to a custard like consistency dependent on whether evaporation is contained or sedimentation rate is important. This factor also allows a much wider range of temperature of operation of the MRS and a better tolerance of dirt, water, air, etc. since the voltages involved are much lower though the field intensity will be $\sim 1T$ at $250kAm^{-1}$ for $\hat{\tau}_e \sim 100kPa$ [8].

Effective conventional mixtures of both EMF and MRS have various amounts of dispersed phase. The speed of the structuring in both media is fast $\sim 1ms$ once the excitation is applied but (in ERF), like the current (doubles for every $\sim 5°C$ from typically 1 to $2mAcm^{-2}$ for $\hat{\tau}_e \sim 3kPa$) can be very temperature dependent. Often the fields need to be forced. The effect of the wall material and condition is not fully resolved.

Figure 8.10 shows ERF field forcing with "shorting" — charge dumping being a particular feature of electrical supply circuits — this is the equivalent of sucking the air out of a motor car tyre to deflate it rather than letting it deflate naturally. In this respect the excitation circuit supply is akin to the catch principle, referring to supply and earth alternatively via a solid state switch.

Whilst there is little hysteresis in a traditional sense, especially in MRS, so long as the fluid is flowing what seems to be particle structure orientation or settlement has been observed at near $\dot{\gamma} \to 0$, when particles are non-

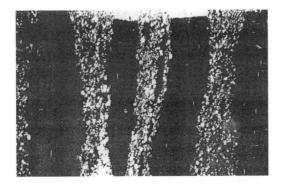

Fig. 8.8 Corn starch particle columns in stationary, excited ERF; No. 30 motor oil, mass oil to starch 1:0.05 at $2.5kVmm^{-1}$, $63\times$ magnification, polarised light [after D. G. Frood].

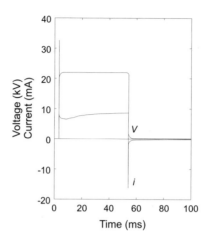

Fig. 8.9   Field forcing of an ERF clutch [after R. C. Tozer].

circular. Microstructuring effects? Particles which are circular at the start of their service may not remain so — likewise particle size.

When field induced structure breaks it tends to do so suddenly. Meanwhile in any hysteretic region the effect of excitation change is small; this aspect of the flow/fluid characteristics reinforces the digital aspect of device performance — on/off rapid switching of the excitation is called for in the strategy to overcome nonlinearities and thermal effects and avoid stabilised power supplies. When shearing is always present repeatability is very good.

The choice of particle size is not so much a factor since it is restricted by the avoidance of Brownian motion, sedimentation and the requirement that the mixtures do not cause blockage. This factor is also influenced by the size of the pole separation (for flux supply reasons to say 1.5–0.5mm) which fixes particle size d at -1 to $+10\mu$m. A mixture of particle sizes may be used to produce a low zero excitation viscosity.

8.6   Performance Prediction

The area of electrostructured fluids is multidisciplinary with geometric forms exerting strong influence on fluid velocity and electrical field distributions. Also, the fluids are barely continua with the scenario switching from macro to micro dependant on the specific problem.

Since a large amount of work may be done in a small space heat generation can lead to the presence of significant temperature gradients in the fluid or design is affected by the avoidance of them. Hence the normal controller/flow choice may be two-dimensional and non-adiabatic. Shear rates from zero to tens of thousands per second can be present in a single device. At the present juncture simultaneous satisfaction of thermal-electrical-dynamic and design/production zones of influence make it difficult to arrive at an optimum machine arrangement. Steps are being taken to overcome this situation by use of computational fluid dynamics (CFD) packages which can be altered to suit ESF characteristics and to include heat transfer with two-dimensional torque/force and directions of flow. These works are in their infancy since the CFD and computer means are still maturing. Meanwhile more patch by patch methods are called for. Where testing and modelling has been possible it is surprising how closely the theoretical prediction and experimental performance conform.

In all of the above the steady state has been implied but the smart machine concept is based on the rapid provision of computer governed generation of force, motion or displacement patterns against time. Most unsteady situations are only really predictable by using CFD which is presently being introduced to include the variation of geometry and/or flow rate with time. Unlike the steady state some discrepancy separates experiment, traditional analysis (where and when it can be applied) and experiment. Much remains to be uncovered before any substantial fast prototyping methodology is in place in a comprehensive sense.

For some time it was argued that because of controller geometrical effects on structure that shear stress as a function of excitation field was totally device dependant. This situation probably arose from the fact that the data achieved from valve flow or in shear flow experiments had been derived on the basis that the device and fluid relationships were the same in each case and that Newtonian fluid relationships could be assumed. These assumptions are, in the limit, erroneous since severe plugs can be postulated via a Bingham plastic, isothermal analysis approach and these affect the flow/pressure, etc. performance. However, the approach was rationalised by using the appropriate Buckingham equation (for laminar flow of a Bingham plastic) to derive $\dot{\gamma}$ and τ_e and by plotting the results non dimensionally.

A high speed viscometer test (torque versus speed) can give a close approximation of shear stress versus shear rate (τ_e vs. $\dot{\gamma}$) since plugs are only formed at very low speeds — provided temperature effects can be con-

tained. From such data can be crudely predicted the performance of valve
or clutch controllers and indeed combinations such as the hydrodynamic
bearing situation.

For the MRS in particular the experimental valve characteristic looks
almost like the τ_e vs. $\dot{\gamma}$ performance (Fig. 8.5(a)). Hence, Newtonian flow
prediction can be achieved for $EX = 0$ by the use of the basic μ_0 whilst
the yield effects can be added independently of $\dot{\gamma}$ viz:

$$T = 2\pi r^2 l \tau, \tag{8.6}$$

where $\tau = \tau_e + \tau_0$ and $\tau_0 = \mu_0 \dot{\gamma}$, also

$$\dot{q} = -\frac{bh^3}{12\mu_0} \frac{\Delta p_0}{l} \tag{8.7}$$

and

$$\Delta p_e = 2\tau_e \frac{l}{h} \tag{8.8}$$

for clutch and valve type controllers respectively.

Great care must be exercised in defining the various terms. For unsteady
flow *per se* compressibility effects can obscure the time response of the
structuring effect in valve or pressurising type flows. Not only can the
velocity profile change but, unless great care is exercised the driving element
can slow down or the excitation mechanism be regulated. There will be a
lapse in time before the overall response develops fully. In general the
excitation should be designed to act ten times faster than the shear stress
rise time expected and this should itself be ten times faster than the time
allocated for the change in motion to take place.

Electrostructured Fluids Rough Sizing Examples

Consider the apparatus of Fig. 8.4 to be impacted with 15,000J of energy
E which has to be absorbed over a 0.1m stroke δ.

Average force on piston rod: $\dfrac{E}{\delta} = \dfrac{15,000}{0.1} = 150,000\text{N}.$ (8.9)

Pressure drop over valve: $\Delta p = \dfrac{150,000}{A}.$ (8.10)

Taking the active area of the piston $A = 0.01\mathrm{m}^2$ then

$$\Delta p = \frac{150,000}{0.01} = 150\text{bar}.$$

For a maximum yield stress $\hat{\tau}_e = 100\text{kPa}$ so, from Eq. 8.8, with $l = 100\text{mm}$ and $h = 0.5\text{mm}$:

$$\Delta \hat{p}_e = 2\hat{\tau}_e \frac{l}{h} = 2 \times 100 \times 10^3 \times 200 = 400\text{bar}. \tag{8.12}$$

If the velocity of the piston rod $U = 10\text{ms}^{-1}$ then

$$\text{Event time:} \quad \Delta t = \frac{\delta}{U} = \frac{0.1}{10} = 10\text{ms}. \tag{8.13}$$

$$\text{Valve flow rate:} \quad \dot{q} = UA = 10 \times 0.01 = 0.1\text{m}^3\text{s}^{-1}. \tag{8.14}$$

For unexcited laminar flow

$$\dot{q} = -\frac{bh^3}{12\mu} \frac{\Delta p_0}{l}, \tag{8.15}$$

so assuming a Newtonian fluid and taking the unexcited dynamic viscosity $\mu_0 = 100\text{mPas}$ and valve width $b = 1\text{m}$ (nominally) then

$$\Delta p_0 \sim \frac{0.1 \times 12 \times 0.1 \times 0.1}{1 \times (0.5 \times 10^{-3})^3} \rightarrow 1,000\text{bar}.$$

This must be brought down so as to increase the excited/unexcited control ratio. At the outset it must be recognised that only a magnetorheological fluid (MRF) can do this job. Such a fluid would have nominally a high viscosity $\sim 300\text{mPas}$ but, the pole (gap) separation h could be quite large — say 1.5mm. Ratioing the above answers:

$$\Delta p_0 \sim 1,000 \times b \times \frac{3}{27} \rightarrow 100\text{bar}$$

and

$$\Delta p_e \sim 125\text{bar}.$$

If a control ratio $\frac{\Delta p_e}{\Delta p_0}$ of 1 is acceptable then this may suffice. Higher pressures could be used to reduce the piston area but the unsteady state, when crudely assessed, contradicts any perceived advantage:

$$\text{Bulk modulus of MRF:} \quad \beta = \frac{v\Delta p}{\Delta v} \rightarrow 2 \times 10^6 \text{Nm}^{-2}, \tag{8.16}$$

where v is the volume of liquid compressed by amount Δv and

$$\Delta v = \frac{v\Delta p}{\beta} = UA\Delta t. \tag{8.17}$$

Given that $F = A\Delta p$ and $v = AL$ then

$$\Delta t = \frac{FL}{A\beta U} \rightarrow \frac{150,000 \times 0.1}{0.01 \times 2 \times 10^9 \times 10} \sim 0.1\text{ms}. \tag{8.18}$$

Generally speaking for rapid response the stroke must be kept down and the area kept high. A stroke of 1m and an area of 0.01m^2 would require a 10ms 'ideal' or theoretical time constant (time for fluid compression if valve is locked).

The full solution would involve a time step by step solution which matches the origin of the force F and velocity U to the valve equation. This could be done by assuming constant temperature but, a steady state thermal assessment of the cylinder alone shows

$$E = mc_v\Delta T. \tag{8.19}$$

Using the c_v of the oil alone, for a 0.1m stroke:

$$150,000 \sim 0.75 \times 2 \times 10^3 \times 0.1 \times 0.01 \times 4,000 \times \Delta T,$$

which means

$$\Delta T \sim 25°\text{C}.$$

This assumes a 25% solid concentration in an oil of $c_v = 2\text{kJkg}^{-1}°\text{C}^{-1}$, giving an average density of $\sim 4000\text{kgmm}^{-3}$. The magnetic circuit design is a specialised matter and probably involves field forcing. It is not attempted here and any heating effect of it on ΔT is ignored.

The same cannot be claimed for the second example — an ERF driven cylindrical Couette type clutch. Here, often, only a small amount of fluid will be sealed within the clutch. The fluid temperature rise can be quite high since the conductance of ERF increases dramatically with temperature — electrical heating can be equal to the heat generated by the shear stress between shearing plates. Heat transfer calculations are the order of the day with the speed of rotation of the normally outer driven plate being an important factor.

When the difference in speed between excited clutch plates is non-trivial, plugs will not exist in the fluid film and $\dot{\gamma} \to \frac{\omega R}{h}$. Hence the torque T being transmitted is

$$
\begin{aligned}
T &= AR(\tau_0 + \tau_e) \\
&= 2\pi R^3 \frac{\mu \omega l}{h} + \tau_e 2\pi R^2 l,
\end{aligned}
\tag{8.20}
$$

where ω is the rotational speed, R is the mean radius, A is the plate area and l is the length of a cylindrical clutch. For thermally stable steady operation the heat transfer from the outer case and shaft must balance $\omega(T_0 + T_e)$. Generally, conductance at a given temperature increases with shear rate whilst τ_e falls (a magnetic coil could complicate this matter — in the case of an MRS clutch).

Of particular importance is the torque/inertia ratio $(\frac{T}{I})$. If motion switching is over frequent severe electrical heating can occur. $\frac{T}{I}$ can be otherwise very important since it is likely to fix the desired acceleration and rate change in velocity or displacement profile.

There seems to be little to choose between radial and cylindrical clutches save in their relative heat transfer capability. The amount of heat conducted along the clutch shaft can be a large proportion of the total heat transfer.

The clutch fluid has a linear equivalent stiffness or rigidity modulus of the fluid G to the bulk modulus of the fluid in the valve. However, any deflection of the fluid film, one plate relative to the other, is likely to be insignificant compared with that of the connected load materials: they usually have much larger dimensions than the fully structured film. Also, the time to accelerate the fluid film seems to be $\sim$1ms, even though the velocity profile must change.

Piezo Valve Sizing

CFD packages have recently been applied to the very tricky problem of the design of these valves in unsteady flow situations. It appears that the cost of the high current charging/discharging supply will prove to be the restriction on the frequency of analogue operation of the piezo driven device whilst the bulk of the coil, bias field mechanism and driver mass will limit the application of its megnetostrictive equivalent. In both cases the initiative is practically quite new. Fast operation is the aim, and seems achievable from experiment and CFD predictions to date.

In both magneto and electrostrictive designs any severe hysteresis in the polarising mechanism can be compensated for adequately by the embedding of strain gauges in the stacks and/or standard algorithms developed for the purpose.

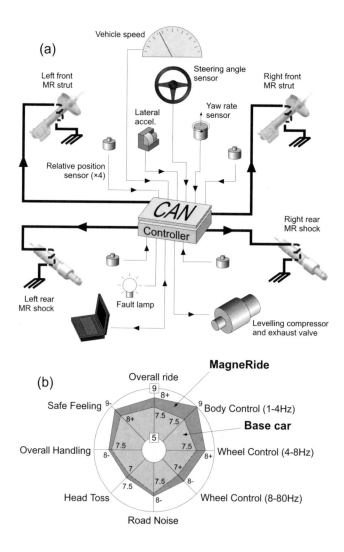

Fig. 8.10 Magneride: (a) system architecture; (b) subjective ratings against the base (standard production) car using a scale of 0–10 (fair: 6, good: 8, excellent: 10).

In terms of simple sizing practice, the axial force on the stack for steady conditions is the product of effective poppet face area times pressure, whilst the flow rate may be determined from the data of Fig. 8.7. For a wider range of openings equivalent data can be obtained from conventional poppet valve test data.

8.7 Applications

Commercial applications of smart fluids devices spearhead the smart technology field and are increasing at a rapid rate. Apart from the balanced gait and stability given to 'above the knee' 'natural walk' prostheses and walking control frames more dynamic applications are in vogue.

Over 5,000 under-seat Lord Corporation Motion Master MR dampers/ suspension systems have collectively travelled millions of miles on class 8 trucks, buses and now agricultural tractors, without undue failure. A competition car (Carrera) MR suspension was tried successfully and in 2002 the Delphi MagneRide system was fitted to that years Cadillac Seville model. The stand alone MagneRide hardware consists of a single pressurised valveless tube charged with N_2 and Lord Corp. MRF — separated by a free piston. The main piston contains the MRF controller and is connected to the vehicle via a piston rod. Added between the vehicle body and all four wheels the cylinders produce a quiet ride plus comfortable suspension and good handling. The systems is said to be three times faster than a fully active model one which it is more bulky. An unexcited parallel flow channel would allow the force to go to zero as motion ceases. The fast response and this cushioning, lead to effective onboard computer controlled real time ride for the first time (Fig. 8.10).

The experience gained via persistence with such high tech projects is leading to studies on vehicle 'steer by wire' systems which incorporate 'feel of road' feedback — another useful property of a hydraulic system. Also, road reading devices could be incorporated.

Civil engineering applications are in place via 30 tonne seismic dampers which respond to stochiastic inputs and employ a parallel or bypass control valve: they are installed into *in situ* chevron braces in buildings in Japan. Simultaneously vibration problems at the Dongting cable stay bridge in Hunan Province, China, are overcome by fitting MR dampers between each cable (different natural frequencies) and the deck (Fig. 8.11).

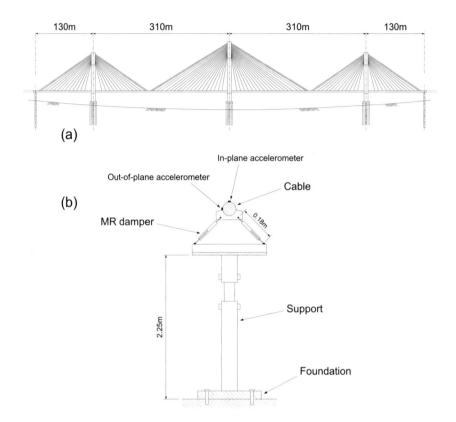

Fig. 8.11 Dongting Lake Bridge: (a) Elevation; (b) Connection of MR damper with cable and deck by use of a supporting pole.

In the former Soviet Union loudspeakers use MR fluids (Fig. 8.12) in the actuation interspace to improve performance and paste type ER fluids of very high yield stress capacity are used to 'stick' non-magnetic objects to grinding machine chucks.

The QED Inc. lens grinder/polisher machine was the first ERF flexible machine commercial success. Diamond dust or similar is added to an MR suspension which can be modulated rapidly to change the 'pressure of the polishing pad', to suit recent lens profile feedback and produce better than hand polished lenses. The lens (Fig. 8.13) spins whilst the 'bearing shaft' delivers fluid to the contact zone after the fashion of the Rayleigh step controller of Fig. 8.4(c).

Further similar developments can be expected, e.g. in shock control and large engine suspensions and, with six companies able to produce smart fluids (MR: Lord, QED, Heat and Mass Transfer Institute, plus New Age Materials; ER: Fluidcon, Smart Technologies and the Asahi Group) this rate of exploitation looks set to continue.

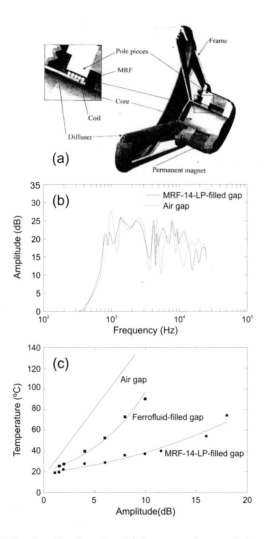

Fig. 8.12   (a) MRS assisted loudspeaker; (b) frequency characteristics of high-frequency loudspeaker; (c) temperature of the voice coil [after HMTI].

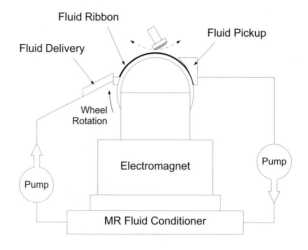

Fig. 8.13   MR polishing system [after QED Inc.].

Inexpensive horizontal axis washing machine MR soaked sponge piston dampers are in the pipeline as are MR plus ER fluids (dual action) and liquid–liquid ESF mixtures of 'true' controllable viscosity. ER inkjet printers are being developed. Control strategies are the object of intense study and tribology mechanisms have come to the forefront recently.

The future may involve further fluid improvements probably from studies of the microstructure and self-assembly of the electrostructured fluids and controller surfaces in action, and exploitation of findings. Already claims of yield stresses of the order of dozens of of kilopascals for ER as well as MR fluids are leaking from several sources. Microstructure manifestation may spell the end for continuum modelling but replacement CFD programmes, e.g. for dissipative particle dynamics, etc., are in development.

Until recently piezo driven valves have been at the pre-prototype stage of development. Nevertheless they show roughly the same speed as a servospool valve and have identified applications in clutch closing in a hydroviscous coupling and, 'rock burst' pit prop piloting. They are presently fitted on the Peugeot 307 (2×10^6 fitted in 2001), common rail IC fuel metering injection systems whereby a degree of management is added to the engine fuel supply (Fig. 8.14). If 1% electrical strains presently being experienced in single piezo crystals can be fully harnessed, small rapid acting, full flow valves could revolutionise the hydraulics field. The feasibility of

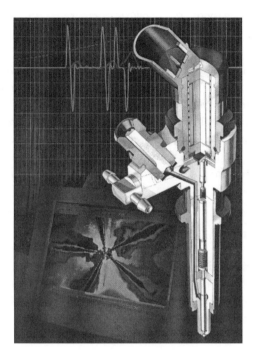

Fig. 8.14 Piezo driven poppet valve embodied in Siemens Common Rail Diesel Fuel
Injector. Several discrete amounts of fuel can be injected during a firing stroke at flexible
electronically directed intervals and durations. Fluid enters valve at high pressure.

control of shape of rate delivery of fuel during one cycle of a diesel engine,
via common rail injection, is a classic of modern smart trends: to trim the
overall system performance with judicious design of the smart element [7]
geometry rather than material property — to give the overall system performance. This produces much lower fuel consumption by the rapid adaption
of the engine to the duty cycle.

Further Reading

(i) Millar Henrie, A. J. (2002) "Magnetorheological Fluids", *Encyclopaedia
of Smart Materials*, Wiley.

(ii) Filisko, F. E. (2002) "Electrorheological Fluids", *Encyclopaedia of Smart
Materials*, Wiley.

(iii) Tao. R. (ed.) (1999) *Proc. 7th Int. Conf. on Electrorheological Fluids, Magnetorheological Suspensions and their Applications, Honolulu, Hawaii.*

(iv) Nakamo, M. and Koyama, K. (ed.) (1997) *Proc. 6th Int. Conf. on Electrorheological Fluids, Magnetorheological Suspensions and their Applications, Yonezawa, Japan.*

(v) Bossis, G. (ed.) (2001) *Proc. 8th Int. Conf. on Electrorheological Fluids, Magnetorheological Suspensions and their Applications, Nice.*

(vi) Bullough, W. A. (1988) "Integrated Electron Hydraulics", *Proc. Australian Bicentennial Int. Congress in Mechanical Engineering, Brisbane,* pp. 18–23.

Acknowledgements

Thanks are due to Asahi, Lord Corporation and the Heat and Mass Transfer Institute, Minsk, for free use of their data and figures; also to Siemens VO, A. Alexandrisis of Delphi Auto and Y. Q. Ni of the Hong Kong Polytechnic University for permission to republish certain figures.

Bibliography

[1] Akio Inoue, Ushio Rhu and Syoji Nishimoro (2001) "Walker with Intelligent Brakes Employing ER Fluid of Liquid Crystalline Polysiloxane", *Proc. 8th Int. Conf. on Electrorheological Fluids, Magnetorheological Suspensions and their Applications, Nice*, pp. 520–525.

[2] Akio Inoue, Yochroh Ide, Shinyungi Maniwa, Hiroyuki Yamada and Hiruyi Uda (1997) "Properties of ER Fluids Comprised of Liquid Crystalline Polymers", *Proc. 7th Int. Conf. on Electrorheological Fluids, Magnetorheological Suspensions and their Applications, Yonezawa, Japan*, pp. 520–525.

[3] Anon. (2000) "Indecisive Material Makes a Big Step Forward", *Design Eng.*, 34–35.

[4] Bullough, W. A. and Peel, D. J. (1986) "Electro-Rheological Oil Hydraulics", *Proc. Japan. Soc. Hydraulics and Pneumatics*, **17**, 520–525.

[5] Stringer, J. D. (1976) *Hydraulic Systems Analysis*, MacMillen.

[6] Strandh, S. (1979) *The History of the Machine*, Bracken Books.

[7] Anon. (2002) "Ceramic Muscle", *Automotive Eng.*, pp. 52–54.

[8] Jolly, M. R., Bender, J. W. and Carlson, J. D. (1998) "Properties of Commercial Magnetorheological Fluids", *Proc. SPIE Int. Symp. on Smart Structures and Materials, San Diego, California*, pp. 1–15.

[9] Alexandrisis, A. (2000) "Magnetorheological Fluid Based Semi-Active Suspension System", *Proc. Euro. Conf. on Vehicle Electronic Systems*, pp. 1–12.

[10] Ko, J. M., Zheng, G., Chen, Z. Q. and Ni, Y. Q. (2002) "Field Vibration Tests of Bridge Stay Cables with Magnetorheological (MR) Dampers", *Proc. SPIE*, **4969**, 1–12.

Bibliography

Chapter 9

Smart Biomaterials — "Out-Smarting" the Body's Defense Systems and Other Advances in Materials for Medicine

Richard van Noort, Paul V. Hatton and David B. Haddow

Centre for Biomaterials and Tissue Engineering,
University of Sheffield,
Claremont Crescent, Sheffield S10 2TA, UK.

9.1 Introduction

Human life expectancy has nearly doubled in the last 150 years and, along with these demographic changes, so has the percentage of those suffering from severe physical disability (Fig. 9.1). The loss of function, whether resulting from injury or disease, is a major health care problem and the option of transplantation of tissues or organs for these patients cannot be met from that source alone due to a lack of availability of suitable donors (www.unos.org). Improved medical devices to restore or replace function are necessary to meet the expectations for a continuation of a 'normal' and active lifestyle through into old age. It is estimated that some 40% of women over the age of 50 will suffer an osteoporotic fracture (www.bonejointdecade.org). Hence mechanical devices and prostheses play an important role in the replacement or repair of deteriorating body parts.

In the context of this article, a *biomaterial* can simply be defined as any material that is used to replace part of a living system or to function in intimate contact with the living system. A more detailed definition was proposed by Williams *et al.* [1] in 1992, who suggested that "a biomaterial is any substance, other than a drug, or a combination of substances, synthetic or natural in origin, which can be used for any period of time, as a whole

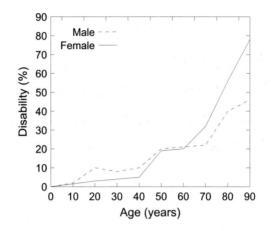

Fig. 9.1   Increasing disability with age for the male and female population.

or as a part of a system which treats, augments, or replaces any tissue, organ, or function of the body". There are many examples of this such as hip and knee joint replacements, heart valve replacements, breast implants and dental filling materials (see Table 9.1). In fact, the chances are that many of the readers of this article will have a biomaterial replacing part of their body. In contrast, a *biological material* is a material of natural origin such as collagen, cellulose or keratin.

The cost of providing replacement parts for the human body has been variously estimated to be in the region of £10 billion. In the UK heart valve replacement accounts for approximately 20% of open heart surgery in National Health Service hospitals with some 7000 artificial heart valve being inserted at a cost to the NHS of £12–15 million (www.ctsnet.org/section/outcome). Although not as dramatic as the life threatening situation associated with a failing heart, the annual cost of materials alone for replacing lost tooth tissue with fillings and prostheses can be conservatively estimated to be in the region of £50 million for the UK alone.

To some degree, all biomaterials used in the production of medical devices to treat human disease may be considered "smart" in that they bring their individual properties to the aid of the surgeon. However, they generally also bring their own intrinsic deficiencies that limit both their usefulness and lifespan. The history of medical devices has, until now, been one of adoption, adaptation and refinement. "Adoption" from other engineering sciences, "adaptation" for medical use and "refinement" in the face of challenges presented by the human body. The degree to which these biomaterials may be considered truly smart or intelligent is subject to debate. This article explains the relationship between properties, design and performance for materials used in medicine, and argues that only now are we truly on the verge of creating smart biomaterials that enhance their own clinical performance by responding to their environment and the living body.

In terms of *"intelligence"*, many biomaterials can be considered to induce an appropriate response when placed in contact with the complex biological environment of the human body. What is considered appropriate or "smart" will vary from device to device. In dentistry there is a tooth-coloured filling material called a glass-ionomer cement, which has been in clinical use since its invention by scientists at the Laboratory of Government Chemist in 1968. This material is smart because it is tooth-coloured and blends in with the adjacent teeth so as to be difficult to differentiate from the natural teeth. More importantly, it also has the capacity to release fluoride such that it provides local protection against the ravages of tooth decay. However, it is still not as "smart" as human tooth enamel that has anisotropic mechanical properties that protect it from excessive mechanical loading, and also has limited capacity for self-repair. Another example of a smart biomaterial is the resorbable suture made from poly (lactic acid)

Table 9.1 Examples of biomaterials and medical devices used to replace or augment human organ function.

Organ or Tissue	Examples
Heart	Cardiac pacemaker, artificial heart valve, total artificial heart
Skin	Suture, tissue adhesives
Eye	Contact lens, intraocular lens
Ear	Artificial stapes, cochlear implant
Bone/joint	Hip joint, bone plate, intramedullary rod, bone cement
Kidney/urinary tract	Kidney dialysis machine, catheter
Skull/brain	Skull fracture repair plate, hydrocephalus shunt
Teeth/oral cavity	Restorative filling materials, permucosal dental implants

or similar soluble polymers. What can be smarter than performing the job intended, and when the period of usefulness has come to an end to simply disappear without any permanent adverse event! The latter is a very important consideration as the biomaterial must be biocompatible, which means that it must not cause any adverse reaction whether local or systemic. Biocompatibility may be defined as the ability of a material to induce/perform with an appropriate host response in a specific application [2]. The response should be such as to ensure the continued safe and effective performance of the material. Furthermore, such materials should not be toxic or have unfavourable properties due either to direct contact or breakdown products. In addition, factors including physical, mechanical properties, design, and surface chemistry should also be considered carefully.

The safety of medical devices is a matter of great concern. Numerous biomaterials have been directly associated with the onset of medical problems following implantation. In some cases this was due to poor material selection or device design, such as the Proplast® temporomandibular joint (TMJ) marketed by the Vitek Corporation in the US. This broke down shortly after surgery, releasing debris into the patient's tissues that frequently resulted in chronic inflammation and pain. Even where the medical device could not be directly related to a reported medical condition, biomaterial manufacturers have found themselves in court. The experience with silicone breast implants was a salutary lesson for companies engaged in the business of developing and marketing of implants. Thousands of women claimed that breast implants inserted during the 1980s and 1990s were the cause of a variety of diseases including breast hardening, shape loss, chronic pain in the muscles and joints, autoimmune problems and other symptoms. While some of the problems were known side effects of breast augmentation, many alleged diseases were difficult or impossible to diagnose. Surgeons suspected that some patients suffered from psychological problems rather than silicone-related medical conditions. Despite the lack of scientific evidence, Dow Corning were forced to settle many of the claims for compensation. The cost of litigation and damages ran into billions of dollars, and Dow Corning had to file for bankruptcy protection. This case raised many wider concerns regarding the protection of medical device manufacturers from ill-defined problems reported by the recipients of their products. At one time, the entire medical silicone rubber industry was threatened by this litigation and almost collapsed. Medical silicone rubbers are used in almost every branch of medicine (from catheters to hydrocephalus shunts), and its loss would have been disastrous. This nightmare scenario was only narrowly averted. One more positive outcome of this and similar cases was the development of improved regulations for medical device manufacture.

All medical devices now have to undergo rigorous safety testing procedures before they can be marketed and used on patients. The process of CE marking was introduced during the 1990s for all medical devices to be used in the European Community. It is aimed at ensuring that medical devices are safe and fit for their intended purpose. There is equivalent legislation in the US provided by the Food and Drug Administration (FDA). However, despite the many safety tests carried out, there can never be a guarantee that a biomaterial will be 100% safe, and post-market surveillance is extremely important. In the UK, severe adverse incidents are reported to the Medical Devices Agency, who will then decide what action is necessary (www.medical-devices.gov.uk/mda-aic). Other systems aimed at monitoring the extent of adverse reactions provide supplementary information on the safe use of biomaterials, such as the adverse reaction reporting project for dental materials (www.shef.ac.uk/uni/project/arrp). This relatively new regulatory environment has proved both a handicap and benefit in the development of improved biomaterials and medical devices. It is certainly no bad thing that manufacturers and clinicians have clear guidelines on good practices related to device design, manufacture and use. In addition, when problems occur they are investigated fully by the MDA and appropriate action is demanded. The only difficulty is that new innovations must be more carefully researched before clinical use in real patients. However, this is a small price to pay for increased confidence in our medical devices and biomaterials.

The successful development of new biomaterials in medicine and dentistry depends on a multitude of factors, such as materials properties, design and the biocompatibility of the materials used. This must therefore be based a multidisciplinary approach as it requires an input for materials scientists, engineers, biologists and clinicians. The common thread that unites these individuals is the desire to develop better devices for medical applications. It was not until the middle of the 20th century that this combination of expertise began to be harnessed in a coordinated way such that now we have deliberate collaborations across the various disciplines.

9.2   Dumb Biomaterials — The First Generation

The idea of using natural or man-made materials to replace parts of the human body and replace function has been around for many centuries. However, it was the advent of anesthesia, aseptic surgical techniques and the discovery of the benefits of x-rays during the latter part of the 19th century that paved the way for the development of novel biomaterials. Before then, the use of biomaterials had not proved particularly successful due to the problems of infection and persistent drainage was the rule. Bone resorption and extrusion of the implant were common problems [2]. Even now infection continues to be a common problem as it is possible for blood borne bacteria to attach themselves to an implant and form a biofilm, which protects the bacteria from attack from the body's natural defense system or the administration of antibiotics. This is analogous to the formation of plaque on teeth, which prevents the saliva from neutralising the acid generated by the bacteria as these digest carbohydrates (sugars), resulting in the dissolution of the enamel surface of the tooth.

In the early 1900s bone plates were beginning to be used as a means of providing fixation of fractures of the long bones. This move towards internal prostheses stemmed from a knowledge that foreign bodies such as bullets or shrapnel may be tolerated by the body for many years [2]. Unfortunately, problems arose due to poor mechanical design or the wrong choice of materials. A material such as vanadium steel was found to corrode rapidly in the hostile environment of the human body and the corrosion products were found adversely to affect the would healing process. The situation improved with the introduction of stainless steels and cobalt–chromium alloys and much better results were achieved for fracture fixation.

Although the first hip joint replacement was attempted in 1938, it was not until the introduction of the cemented hip prosthesis by Charnley in 1958 that the first consistent results were produced (Fig. 9.2). Polymethyl methacrylate (PMMA) was used in these early cases for the cementation of the hip joint. The suggestion for using PMMA was made by Dr. Dennis Smith, a dentist working in Manchester University. PMMA is widely used for the construction of acrylic dentures and has handling characteristics that makes it suitable as the cement for holding the hip joint in place. Also, it was known from war time experiences in World War II that air crew who had suffered injuries, where fragments of PMMA had become embedded in their bodies, did not appear to suffer any chronic adverse

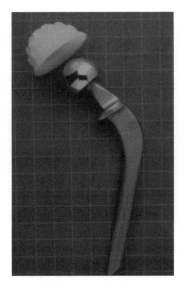

Fig. 9.2   Stainless steel hip joint and high density polyethylene acetabular cup.

reactions. On the basis of this observation PMMA was being explored in the 1940s for corneal replacement. The combination of stainless hip joints and PMMA bone cement has since proved very effective and to this day most hip replacements are still cemented with PMMA.

A few years after Charnley's success, another orthopaedic surgeon by the name of Bränemark [3] made a fortuitous discovery regarding the biological response to titanium. He had developed a rabbit model to study blood flow through the peripheral blood vessels by creating a glass window set in a ring of titanium. When he wanted to reuse these windows he found that they were extremely difficult to remove because the cartilage of the ear had become strongly attached to the alloy. Further experiments in bone showed that the biological response to titanium was quite different from that experienced with stainless steel or cobalt–chromium alloys. For stainless steel and Co–Cr alloys, the body's response tends to be a process of fibrous encapsulation of the implant. This is a defense mechanism against the invasion of a foreign body and is commonly seen with a lot of synthetic materials. In contrast, titanium was found to stimulate an intimate relationship between the implant and the bone, such that the implant becomes embedded in the bone. Bränemark called this type of response

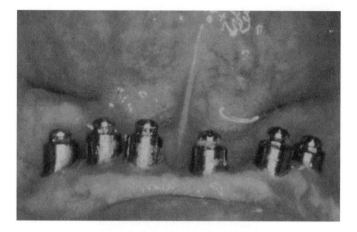

Fig. 9.3   Dental implants made from commercially pure titanium protruding from the soft tissue.

osseintegration, distinguished from other biological responses by the close adaptation of bone to the surface of the implant without the formation of a fibrous capsule. This observation has lead to the development of cementless fixation of hip prostheses using Ti_6Al_4V alloys and the development of the highly successful permucosal dental implants (often referred to as screw-in teeth) using commercially pure titanium (Fig. 9.3)

Simultaneously in the cardiovascular field it was also noticed that some materials performed better in contact with blood than others. One of the body's primary defense mechanisms when injured is for a blood clot to form so as to prevent further bleeding from the damaged blood vessels. In the same way, when blood comes in contact with a foreign object it sets off the clotting process, since we have not yet discovered a way of inserting implants without causing local tissue damage. This has proved a serious hurdle in the development of blood contacting devices such as the artificial heart, blood vessels, oxygenators and kidney dialysis machines.

Gradually there was a realisation that different materials can induce widely different responses in the biological environment. Therefore, in order to overcome the problems encountered in the use of synthetic materials, it was necessary to develop smart *biomaterials*, that is "materials that actively engage with the biological environment in such a way as to achieve the desired result".

9.3 Planning and Refinement — Second Generation Biomaterials

While a number of useful first generation biomaterials were discovered, it was apparent that they often suffered limitations in more demanding applications, especially in the longer term. In load bearing implants, it is very important that there is an appropriate transfer of stress between the implant and the bone. Bone is a very smart material. If bone is not subjected to stress it will resorb on the basis that the message it receives is that it is no longer needed if not stressed, a process known as '*stress shielding*'. It is a problem that astronauts suffer when spending a lot of time in the weightless environment of space. In contrast, where bone is highly stressed it may try to compensate by laying down additional bone so as to reduce the stress. The callus that forms during the healing of a fractured bone is an example of this process. However, if too highly stressed then the bone will recede in order to avoid these highly damaging stress levels. In dentistry, this smart behaviour by bone is utilised to move teeth around, which may be for aesthetic or functional reasons. It forms the basis of the clinical discipline of orthodontics.

In the case of hip prostheses, the design and stiffness characteristics of the materials used are widely different from the natural bone being replaced (see Table 9.2). The high success rate with the early hip joints can be partly attributed to the age of the patients, who were typically older than 65 years of age. Such patients are not especially active and thus the stresses to which the implant and the surrounding bone are subjected are not has high as with younger more active patients. Yet, there is considerable demand for joint replacements in the young and physically active patient. This places

Table 9.2 Comparison of the mechanical properties of cortical bone and some implant materials.

Material	Tensile Strength (MPa)	Elastic Modulus (GPa)	Elongation (%)
Cortical bone	136	18	1
Ti_6Al_4V	860	110	12
Stainless steel	550	200	50
Cast Co–Cr alloy	650	240	10

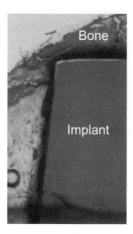

Fig. 9.4   Fibrous encapsulation of a zirconia implant as indicated by the dark band around the implant.

considerable demands on the effective functioning of these implants. Stress shielding can result in aseptic loosening of the implant and is a significant contributor to implant failure. In addition, the acrylic bone cement does not provide a very effective bond to the bone and the local micro-movement that takes place has also been implicated in aseptic loosening of the joint.

The work on titanium showed that some materials may be able to induce a response in bone cells that can be highly beneficial. The cementless fixation made possible with titanium alloys has opened up the opportunity to develop implants that provide a more benign stressing condition, which can improve the longevity of the implant.

Titanium and its alloys are described as osseointegrating in that these materials appear to allow direct bone contact after healing without the intervention of a fibrous, soft tissue layer. Figure 9.4 is an example of the fibrous encapsulation that occurs with zirconia, which contrasts with the close apposition of bone to a composite ceramic made of hydroxyapatite and zirconia (Fig. 9.5).

While some researchers have argued that osseointegration is a feature of the TiO_2 oxide surface of the biomaterial, this is probably only part of the story. The more generally accepted explanation is that it is careful surgery that avoids the death of local bone that provides osseointegration around dental and similar titanium implants.

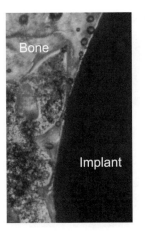

Fig. 9.5   Osseointegrated implant based on a composition of 50% HAp and 50% zirconia.

Some biomaterials are described as having osteoconductive properties, as there is evidence that they stimulate generation of new bone on their surface after implantation into established bony tissue. While titanium is not a truly osteoconductive biomaterial, a number do exist. Since the 1970s, new materials have been developed that enhance bone formation and healing at the site of surgery. Examples of such materials are synthetic hydroxyapatite, bioactive glasses (e.g. Bioglass®) and osteoconductive glass-ceramics (e.g. apatite–wollostentite glass–ceramics). While the exact mechanism is not known, there are two complimentary theories supported by published scientific evidence. The first is that specific proteins are adsorbed from the biological environment that subsequently recruit cells that precursors to bone forming cells (osteoblasts). These adhere to the biomaterial surface and produce bone tissue that assists the overall healing process. In some biomaterials (such as Bioglass), this process is preceded by ion exchange at the implant surface and development of a calcium phosphate layer. This reaction is believed to allow the formation of a direct chemical bond between the Bioglass and the mineral phase of newly formed bone tissue.

9.3.1   Calcium Phosphate Ceramics

The fact that the mineral phase of bone and tooth tissue is composed of calcium phosphate salts, has directed researchers to investigate these ma-

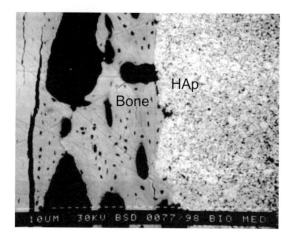

Fig. 9.6   Image of the interface between a HAp implant and bone taken using a scanning electron microscope, showing direct contact between the bone and the implant.

terials as potential candidates for bone substitutes. Although the bone mineral is similar to hydroxyapatite $[Ca_{10}(PO_4)_6(OH)_2]$, it is poorly crystalline in structure and contains a wide range of calcium phosphate phases, including tricalcium phosphate, carbonated apatite and numerous other ionic impurities such as fluoride, magnesium and sodium [4]. Fluorapatite $[Ca_{10}(PO_4)_6(F)_2]$ exhibits the same structure as HAp, the only difference being that the hydroxyl groups have been replaced by fluorine ions. FA is more stable than HAp at high temperatures, and more resistant to acid attack. Tricalcium phosphate $[Ca_3(PO_4)_2]$ can be found in two forms, α-TCP and the more stable β-TCP. Both forms have been reported to exhibit relatively high dissolution properties [5]. These materials can bond with natural bone without fibrous tissue encapsulation. Chemical bonding of calcium phosphate ceramics to bone comprises a partial dissolution of the ceramic surface and the subsequent formation and reprecipitation of CO_3–apatite crystals together with biomolecules of the surrounding tissue fluid [6] (Fig. 9.6).

The development of synthetic calcium phosphates has provided a wide range of materials to be investigated for their bone replacement properties, although their application is limited due to their poor mechanical performance. A major application of HAp is in low stress bearing situations such as alveolar ridge augmentation [7; 8] (Fig. 9.7).

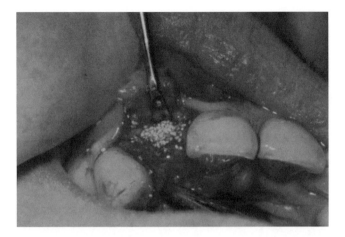

Fig. 9.7   Localised alveolar ridge augmentation using HAp granules.

One way of overcoming this limitation is to apply the calcium phosphate ceramic as a coating on the surface of a metallic implant. This concept of utilising materials that induce the formation of new bone has been widely explored as a means of coating the surface of implants. One technique is to use plasma spraying [9] with smart biomaterials such as HAp that speed up the process of osseointegration. In effect, the biological response is governed by the surface coating rather than the bulk material underneath. In this situation the bulk material can provide the strength while the coating provides the appropriate biological response (Fig. 9.8).

These approaches to developing implant materials with a smart response in the biological environment, can still be considered to be based on empirical observation. In order to control and predict the biological response to a biomaterial it is necessary to have a much more detailed understanding of the material–tissue interface interactions. Fortunately, at the same time as materials scientists were exploring new materials, the biological sciences were developing at a rapid pace.

9.3.2   Bioactive Glasses

In bioactive glasses, the composition is designed such that the surface undergoes a selective chemical reaction with the physiological environment, resulting in a chemical bond between tissue and the implant surface [10].

Fig. 9.8   Plasma sprayed surface coating of HAp on the bone contacting surface of a metal acetabular cup.

The bonded interface protects the implant material from further deterioration with time. The first man-made materials, which were found to bond to living bone, were glasses in the system Na_2O-CaO-SiO_2-P_2O_5. These were discovered by Hench *et al.* [11] in the early 1970s and named bioglasses. The bioglass material seems to enhance bone–implant interface development and bonding when implanted in bone. When bioglasses are subjected to an aqueous environment, calcium and phosphate ions leach from the implant bulk to form a calcium phosphate-rich layer on the implant surface, which is believed to impart the bone bonding ability to the bioglasses. Biological processes, such as collagen interdigitation, have been described by Hench *et al.* [12]. Since the discovery of bioactive glasses in early 1970s, various kinds of glasses and glass-ceramics have also been found to bond to living bone, and many are now clinically used such as an artificial middle ear bone [11] and alveolar ridge maintenance implants [13; 14]. Bioactive glasses are also used as artificial vertebrae, intervertebral spacers, iliac crest prostheses, and granules for bone defect fillers [11; 12; 13]. As with the calcium phosphate ceramics their application is limited to low stress bearing conditions due to poor mechanical properties.

9.4 Smart Surfaces Tailored for Specific Applications — Third Generation Biomaterials

9.4.1 *Materials–Tissue Interface*

When a biomaterial is implanted in the body, the site is invaded by macrophages. If these cells are able to engulf the implant then the cells will attempt to break down the offending object, a process known as phagocytosis. However, if the foreign object is too large then the macrophages will adhere to the surface the implant. The result of this is that a foreign body reaction is initiated, which manifests itself as an encapsulation of the implant by collagenous connective tissue. These macroscopic effects are regulated by nano-scale processes and the surface properties of the biomaterial have a major influence on the type of response produced as it is the surface of the biomaterial that first comes in contact with the body [15].

If the uptake of macrophages and subsequent encapsulation is to be avoided then this is possible either by a judicious choice of the biomaterial based on past experience (e.g. Ti), or by modification of the material's surface chemistry. The former approach is intrinsically very limiting as there are few materials that elicit the type of biological response desired and also the nature of these responses is very limited. The latter approach is in effect a means of 'out-smarting' the body's natural defense mechanism by hiding the bulk material, which may elicit an inappropriate biological reaction, behind a surface layer tailored to give the desired response.

The behaviour of cells in organs and tissues depends on cell–surface interactions both with molecules on the surface of other cells, the surface of the substrate and with the extracellular matrix [16]. Hence the success or failure of tissue engineered devices is dependent on the complex interplay between cells and materials. The response of cells to natural or synthetic materials is therefore crucial in developing laboratory-based strategies for wound care and healing. These cell–surface interactions influence and control cell physiology, such as adhesion, spreading, migration, proliferation and differentiation. When a biomaterial is implanted the first thing that takes place is the adsorption of proteins, such as fibronectin, laminin, vitronectin and collagen, or a variety of cell adhesion molecules (CAMs), from the extracellular matrix. From the work of cell biologists we have learnt that a cell surface is covered by an array of adhesion receptors, which have been identified as immunoglobulins, selectins and integrins. The proteins interact with these receptors on the cell surface and bind the cell to its extracellualar matrix (ECM). The ECM acts as a soup with a wide variety of ingredients such as cytokines and growth factors and orchestrates the whole process.

It is therefore important that the right sort of proteins are adsorbed onto the surface of the implant since these will determine the biological response and thus eventually the fate of the implant. For instance, when a surface is contacted with blood and the response is the formation of a blood clot then such a material would be inappropriate for applications as haemodialysers, artificial blood vessels or heart valves. Similarly the surface a contact lens needs to be easily wetted by tear fluid as it will otherwise damage the cornea. Lack of adhesion between a dental implant and the mucosa would cause infection due to the infiltration of bacteria from the oral environment into the gap.

It is unlikely that 'off-the-shelf' materials will be able to provide the variety of different cell-surface interactions desired. Hence, in order for biomaterials to be accepted by the body, it is vital that the implant surface is *designed* so as to promote appropriate and highly selective interactions with its biological environment. The nature of these interactions will be governed by the application of the biomaterial. Hence a material that may have excellent properties for providing structural support, but does not induce the appropriate biological response, can be made to do so by coating it with a surface layer with a chemistry that is tailored to suit a particular application.

From an understanding of the factors that control cell–surface interactions it is possible to devise methods of modifying the surface characteristics of biomaterials that will have the effect of changing the local biological response. One approach involves modification of the surface chemistry of the biomaterial in order to regulate the type of proteins adsorbed onto the surface, another other is to create highly selective interactions by grafting biologically active molecules onto the surface of the biomaterial. The manner of such surface modifications will be dictated by the nature of the response required. For example, in devices that will be exposed to a bacterial flora such as a synthetic voice box, cell attachment is to be discouraged whereas for orthopaedic implants cell attachment is beneficial as long as these are the right cells and they behave in an appropriate manner.

9.4.2   *Functionalised Surfaces*

Cell–surface interactions have been investigated on ion-exchange materials (charged polymers) [17], self assembled monolayers (SAMs) [18] and plasma polymers [19].

SAMs formed by the adsorption of chemical chains onto gold surfaces consist of chemically robust, ordered arrays of organic molecules that have shown considerable promise as model systems for investigating cell–material interactions. By selecting the end group of the adsorbed chain it is possible to tailor the chemistry of the SAM surface-making SAMs ideal for fundamental studies of protein adsorbtion and cell adhesion. U sing SAMs as model substrates will increase knowledge about cellular interactions with surfaces, but their end clinical use is limited because the chemical chains are adsorbed onto gold layers on glass substrates. Plasma polymers, by contrast, have the potential to treat a range of materials for essential use in patient care.

Functionalised surfaces are able to influence cell behaviour by controlling the nature of the protein layer which spontaneously adheres to those surfaces in the biological environment. The cells are able to sense the proteins using receptors and binding sites and in this way the cell behaviour can be modified. Typically, the type of behaviour that tissue engineers find useful to influence or control are cell attachment, cell growth (an increase in cell size), and cell proliferation (an increase in cell number). For example, carboxylic acid functionalised surfaces can control skin cell attachment and proliferation, whilst surfaces containing nitrogen can influence how nerve cells respond.

An interesting feature of some plasma polymer surfaces is that they can allow cell release at an appropriate site. This has been a considerable challenge for researchers — if a cell is happy to attach to a surface why should it then want to leave that surface? Early studies on acid functionalised plasma polymers show that the surfaces encourage skin cell attachment, but at the site of a wound such as a diabetic foot ulcer, the surface degrades allowing cells to be released into the wound site to kick-start the healing process. This technology could bring the concept of a "living bandage" to the wound healing market.

By using masks to deposit plasma polymers with different functional groups onto a substrate, finely detailed chemical patterns (Fig. 9.9) can be created and imaged using time of flight secondary ion mass spectrometry

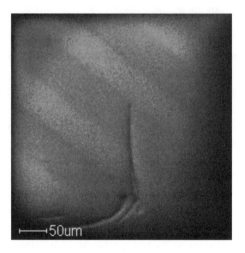

Fig. 9.9   ToF–SIMS image of acrylic acid stripes (m/z 16 + 17, dark) on a plasma polymerised 1,7-octadiene background (m/z 12 + 13, light).

(ToF SIMS). Such patterns possess unique chemical properties with regard to hydrophobicity/hydrophilicity and cell and protein binding (Fig. 9.10). Spatial control of surface chemistry is an essential requirement in emerging technologies such as tissue engineering. For instance, nitrogen containing

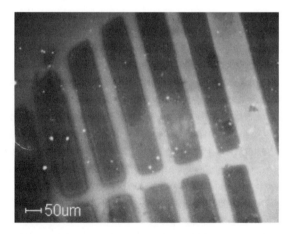

Fig. 9.10   FITC–IgG immobilised on allyl amine/1,7-octadiene plasma patterns — 50µm IgG 'bars' with part of the grid rim.

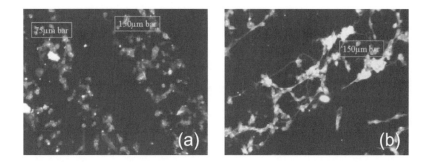

Fig. 9.11 Actin stained cerebellar granule neurons grown on patterned coverslips for (a) 5 hours, showing patterned attachment; (b) 1 day, showing patterning of neurite out-growth.

plasma polymers are able to support nerve cell attachment and process (neurite) growth. The influence of surface chemistry over cell attachment can be used in the spatial control of attachment. The ability to produce spatially ordered nerve cell cultures has been used to assess the directional effects of electrical stimulation on neurons [20]. Such patterns also provide a way to investigate neuronal interactions in a network by growing the neurons in a confined pattern (Fig. 9.11). There are possibilities for use in neuronal and axonal regeneration, by providing neurons with an attractive pathway to extend processes along.

Surface patterning using plasma polymers can allow an object of complex 3-D geometry to be patterned [21]. One of the current challenges with 3-D scaffolds is to encourage cells to grow into the entire scaffold and not merely attach at the first available surface, usually the outer edges. By providing patterned pathways within the scaffolds to promote cell migration through the structure, this problem could be overcome.

9.4.3 *Biologically Modified Surfaces*

The term 'biological modification' implies altering a material's performance by covalently coupling to the material's surface a biologically relevant and active molecule that the tissue surrounding the material recognises through a cellular or biomolecular pathway [22]. Various models that mimic basement membranes, cell surfaces and extracellular matrix have been developed to study the interaction of cells with substrates and how these in-

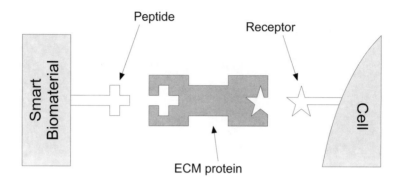

Fig. 9.12 Schematic of a smart biomaterial with a peptide grafted onto its surface so as to attract the desired protein from the extra cellular matrix, which in turn attracts the appropriate cell attachment.

fluence cell behaviour. Such models involve the immobilisation of biologically active ligands of natural or synthetic origin onto various substrates to produce chemically defined bioactive surfaces. These ligands include cell–membrane receptor fragments, antibodies, adhesion peptides, enzymes, adhesive carbohydrates, lectins, membrane lipids and glycosaminoglycan matrix components [23].

The process of grafting ligands onto the surface of the substrate involves selection of a suitable substrate chemistry, which may be modified using such techniques as plasma polymerisation describes above. Such a surface provides highly reactive species for binding the bioactive ligand onto the surface (Fig. 9.12). The choice of ligand depends on the desired cellular response.

9.4.3.1 *Bacterial Adhesion*

Despite the use of aseptic techniques, bacterial infection is a frequent complication associated with implantable devices. The plasma protein fibrinogen is readily adsorbed onto most surfaces and has been shown to mediate the adhesion of *Staphylococcus aureus*. This pathogen is associated with a large number of implant-related infections. Baumgartner *et al.* [24] have explored a series of phosphonated polyurethanes as a means of reducing bacterial adhesion. In their work polyurethanes were synthesised, which directly incorporated a phosphorylcholine ligand into the backbone of the polyurethane. Glycerophosphorylcholine was used as the chain extender in a material composed of methylene diphenylene diisocyanate and poly(tetremethylene) oxide. They found a reduction in bacterial adhesion to this material.

9.4.3.2   *Bone Bonding*

Bone cells have cell surface receptors (integrins) that bind readily to the amino acid sequence arginine–glycine–aspartate (RGD) domain on proteins such as fibronectin and vitronectin in the extracellular matrix. Rather than modifying the functional chemistry of the substrate to encourage selective protein adsorption, short chain peptides containing these cell binding sequences can be grafted onto biomaterials, thus mimicking the more complex proteins such as fibronectin [25]. Various studies using this approach have shown that cell adhesion can be improved significantly using these RGD sequences [22; 26; 27].

9.4.3.3   *Blood Compatible Surfaces*

Most of the currently used artificial blood vessels are made from expanded poly(tetrafluoroethylene) or woven polyester. For clinical situations where there is high blood flow and low resistance (e.g. aortic bifurcation) these materials have proved clinically acceptable. However, in situation of low blood flow, which in effect means blood vessels of less than 4mm diameter, the grafts readily get blocked and the patency drops to only 10% after 8 years [28].

The interaction of blood cells (leucocytes) with endothelial cells, the cells lining the wall of the blood vessel, is highly complex and involves selectin-mediated rolling and integrin-mediated adhesion (Fig. 9.13). For this reason, one of the potentially more effective methods of making a blood-contacting surface haemocompatible is to create a surface on the biomaterial that encourages full coverage with a layer of endonthelial cells. This is possible by grafting cell-adhesive proteins such as fibronectin or cell-adhesive oligopeptides (RGDs) onto surfaces to enhance the adhesion of endothelial cells. An example of this is the carbodiimide method. This system employs a blend of poly(carbonate urethane) and poly(hydroxybutuyl acrylate) [29]. A bifunctional spacer is then applied (carboxylic dichloride) such that carboxylic acid groups are obtained, creating a surface ready for

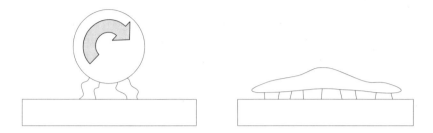

Fig. 9.13   Selectin-mediated rolling and intergin-mediated adhesion of a cell to a substrate.

peptide immobilisation. The next stage is the formation of a carbodiimide–carboxyl adduct and finally the carbodiimide group is displaced by a nucleophilic amine group, resulting in a stable and covalently bonded peptide (Fig. 9.14).

9.5   Really Smart Biomaterials — The Next Generation

In the previous sections, we have seen that injury and disease frequently result in loss of living tissue or function, and that many clinical treatments are based on replacement of lost tissue or restoration of function using synthetic biomaterials and medical devices. In an effort to improve clinical results, increasing efforts were directed at understanding tissue–material interaction and developing biomaterials with improved clinical performance (so called "bioactive" or second generation biomaterials). However, the best "material" for any specific function of the human body is healthy

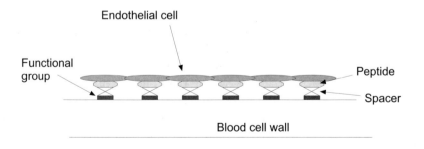

Fig. 9.14   Schematic of the surface of an artificial blood vessel [28]).

living tissue. This fact has given rise to a new philosophy in the field of biomaterials, that of tissue engineering. Tissue engineering has been defined as "the use of biological and engineering principles to construct functional tissues to replace or supplement diseased or defective body parts" (Yorkshire Biomaterials Network, Tissue Engineering Group, August 2000). In practical terms, the idea is to grow live cells on a biomaterial scaffold in the presence of growth medium and biologically active molecules. This combination of live cells (and any extracellular matrix they produce) with the scaffold is subsequently implanted into the site of need in the body as a cell–biomaterial construct. It is likely that tissue engineering will remain closely linked to the field of biomaterials due to this need for a cell "carrier" or artificial scaffold in most applications. The term "substitution medicine" has been coined by Clemens van Blitterswijk (IsoTis BV, Netherlands) to cover the whole range of treatments based on both biomaterials in medical devices and tissue engineering. Tissue engineering itself is arguably one of the fastest developing research fields in the world (Time magazine also places tissue engineers at the top of its "Hottest Jobs of the Future" table). In part, this is due to convergence of scientific disciplines, where advances in biology, chemistry and materials science have come together to make a tissue engineering approach to treatment of disease a viable proposition. Growth has also been fuelled by the increasing healthcare needs of an ageing population, coupled with tremendous expectation of the potential impact of tissue engineering on treatment of disease.

Until now, the biomaterial scaffold was commonly made from an existing bioresorbable material such as PLLA. A degradable biomaterial is considered ideal, as it will most likely be totally resorbed by the body after implantation. While useful, these biomaterials have never been truly optimised for tissue engineering applications, and they may not be perfect because some of their degradation products may inhibit cell growth and differentiation in culture. Research is therefore being carried out novel biomaterials for specific applications in tissue engineering. One area of considerable promise is that of biologically functionalised scaffolds, where the surface presented by the biomaterial provides information to the cells that attach themselves. This information might simply determine where cells should and should not adhere, or it may determine orientation or differentiation. It is anticipated that current research will ultimately provide a range of designer scaffolds for specific tissue engineering applications in the future. The modifications will most likely include elements of the surface engineering described previously, with intelligent degradation and incorporation of biologically active peptides or drugs. While these advances are some way off, there are contemporary examples of intelligent surface modification to control cell adhesion. Okano *et al.* [30] have developed a thermally responsive biomaterial surface using poly(N-ispropylacrylamide) where, below 32°C it is fully hydrated and above 32°C it is hydrophobic. Cell detachment is possible by reducing the temperature below 32°C after growing the cells at 37°C. This may be very useful when developing tissue engineered skin and there is a need to detach the cell sheet from the culture substrate prior to application on the wound bed.

Despite some of the advances above, truly intelligent scaffolds are the materials of the future. It remains difficult to create the right morphological and biochemical environment for directed cell adhesion, growth and differentiation to form specific functional tissues. There is no doubt that, along with further biomaterials research, advances in biological and medical sciences are required to perfect tissue engineering to the point that its use becomes commonplace. Wider issues include cell source (including the use of stem cells) and culture conditions (including the use of mechanical stimulation and bioreactors).

9.6   Conclusions

Traditionally, biomaterials were identified on the basis of their mechanical properties and non-toxicity in the biological environment. Whereas one hundred years ago the field of biomaterials was dominated by the approach of trial and error, nowadays a much more sophisticated approach involving the careful design of biomaterials is taken. The threat of litigation if something goes wrong is sufficient a deterrent to prevent the "cavalier" approach of the early days of biomaterials and medical device development. However, many effective medical devices still suffer from long term failure due to the intrinsic limitations of inert, non-living substances. The development of second generation or so-called "bioactive" materials has been critical, not only in the production of improved medical devices but also as a signpost towards the future strategy that should be adopted. Contemporary research is based upon the creation of a beneficial interaction between host tissues and the biomaterial. Only through this research will we develop long-lasting, effective medical devices and engineered cell–biomaterial constructs to repair or replace failed human tissues and organs. This is not straightforward, as disease generally creates a catabolic environment that challenges successful intervention. Medical device and tissue engineering technologies are still evolving, and biomaterials will undoubtedly continue to play a central role in the treatment of disease. This article predicts that future developments in biomaterials will arise from a close working relationship between biology and materials science. Indeed, it is likely that the new biomaterials for future therapeutic applications may depart quite radically from the biomaterials of the past. New biomaterials will be much smarter in that they will expected to interact with the biological environment in such a way as to modulate physiological functions and actively encourage repair and clinical recovery. The ultimate challenge of tissue repair is the regeneration of the original tissue without the remnants of foreign biomaterials [31].

Bibliography

[1] Williams D. F., Black, J. and Doherty P. J. (1992) "Consensus of 2nd Conference on Definitions in Biomaterials", *Biomaterial–Tissue Interfaces*, **10**, 525.

[2] Burney, F. and Muster, D. (2000) "Aspects of Reconstructive Biomaterials", *MRS Bulletin*, **25**, 15.

[3] Bränemark, R., Ohrnell, L. O., Nilsson, P. and Thomsen, P. (1997) *Biomat.*, **18**, 969.

[4] de Groot, K. (1983) "Ceramics of Calcium Phosphates: Preparation and Properties", *Bioceramics of Calcium Phosphates* (ed. K. de Groot), CRC Press.

[5] Klein, C. P. A. T., de Blieck–Hogervorst, J. M. A., Wolke, J. G. C. and de Groot, K. (1990) *Adv. Biomat.*, **9**, 277.

[6] LeGeros, R., Daculsi, G., Orly, I. and Gregoire, M. (1991) "Substrate Surface Dissolution and Interfacial Biological Mineralization", *The Bone–Biomaterial Interface* (ed. J. E. Davies), Toronto University Press.

[7] Brook, I. M., Graig, G. T. *et. al.* (1987) *Brit. Dent. J.*, **162**, 413,

[8] Frame, J. W., Rout, P. G. J. and Browne, R. M. J. (1987) *Oral Maxillofac. Surg.*, **45**, 771.

[9] Klein, C. P. A. T., Patka, P., van der Lubbe, H. B. M., Wolke, J. G. C. and de Groot, K. (1991) *J. Biomed. Mat. Res.*, **25**, 53.

[10] Hulber, S. F., Hench, L. L., Forbes, D. and Bowman, L. S. (1983) "History of bioceramics", *Ceramics in Surgery*, **17**, 3.

[11] Hench, L. L. and Wilson, J. (1984) *Science*, **226**, 630,

[12] Hench, L. L. (1991) *J. Am. Ceramic Soc.*, **74**, 1487.

[13] Vogel, W. and Holland, W. (1987) *Angew. Chem. Int. UK*, **26**, 527.

[14] Wilson, J. (1985) "Clinical Application of Bioglass", *Glass: Current Issues* (ed. A. F. Wright and J. Dupuy), pp. 662–669.

[15] Ikada, Y. (1994) *Biomat.*, **15**, 725.

[16] Springer, T. A. (1990) *Nature*, **346**, 425,

[17]   Shelton, R. M. and Davies, J. E. (1991) *The Bone–Biomaterial Interface*
       (ed. Davies, J. E.), Toronto University Press.
[18]   Cooper, E., Wiggs, R., Hutt, D. A., Parker, L., Leggett, G. J. and Parker,
       T. L. (1997) *J. Mat. Chem.*, **7**, 435.
[19]   Haddow, D. B., France, R. M., Short, R. D., MacNeil, S., Dawson, R. A.,
       Leggett, G. J. and Cooper, E. (1999) *J. Biomed. Mat. Res.*, **47**, 379.
[20]   Matsuzawa, M. (1994) *Brain Res.*, **47**, 67.
[21]   Bullett, N. A., Short, R. D., Beck, A. J., France, R. M., Cambray–Deakin,
       M., Fletcher, I. W., Douglas, C. W. I., Roberts, A. and Blomfield, C. (2001)
       (preprint).
[22]   Rezania, A. *et. al.* (1997) *J. Biomed. Mat. Res.*, **37**, 9,
[23]   Drumheller and Hubbell (1995) *The Biomedical Engineering Handbook* (ed.
       J. D. Bronzini), CRC Press.
[24]   Baumgartner, J. N. *et. al.* (1997) *Biomat.*, **18**, 831.
[25]   Pierschbacher, M. D. and Ruoslahti, E. (1984) *Nature*, **309**, 30.
[26]   Cook, A. D. *et. al.* (1997) *J. Biomed. Mat. Res.*, **35**, 513.
[27]   Sugawara, T. and Matsuda, T. (1995) *J. Biomed. Mat. Res.*, **29**, 1047,
[28]   Tietze, L. (2000) *MRS Bulletin*, **25**, (1), 33.
[29]   Anderheiden, D. (1992) *J. Mat. Sci. Mat. Med.*, **3**, 1.
[30]   Okano, T. (1995) *Biomat.*, **16**, 297.
[31]   Sittinger, M. (1996) *Biomat.*, **17**, 237.

Chapter 10

Natural Engineering — The Smart Synergy

Julian F. V. Vincent

Centre for Biomimetics and Natural Technologies,
Department of Mechanical Engineering,
University of Bath,
Bath BA2 7AY, UK.

10.1 Introduction

The literature is replete with definitions of smart and intelligent materials. The common thread is that the material (or structure) has to be capable of sensing its surroundings and responding accordingly. Argument arises because there are so many levels at which this functionality can exist, each with its own level of complexity and competence. The general model requires a means of input or sensing, perhaps with transduction, some type of integration or decision-making, and an appropriate output [1]. There is no limitation on the identity of any of these components — the definition is totally functional. At what levels do these types of function occur in biology, and what can we learn from them? This review represents only my own limited views and interpretations. The examples quoted are but a small subset of what is available in nature, but represent areas either where there is current advance in a biomimetic approach, or where there ought to be!

10.2   Intelligent Biomimetics

10.2.1   *Sensory Mechanisms*

The basic unit of life is the cell, representing the level at which response and actuation reside. The most important biological site for transfer and transduction of information is the cell membrane. At this interface external stimuli are converted into chemical messages which can be processed by the chemistry of the cell. Further processing has to be performed by the cell, which then provides the output, again traversing the membrane. The intervention of the cell allows both amplification and integration of inputs. However, biological organisms are so complex that examples which we might implement in some way have to be chosen with care if they are to illustrate or exemplify a problem clearly. As the unit of life the cell also has to perform all the processes of existence, which is not required of simple smart devices such as the bimetallic strip, or photochromic glass. The choice of biomimetic model therefore is probably much better made as a single function found in a larger organism where the constituent cells are able to specialise their functions and the required function can be investigated in isolation and its characteristics defined more clearly. Ultimately the sensitivity of receptors is controlled by quantum effects [2]. This in itself illustrates the sort of approach which is necessary to take in biomimetics, since a sensory system that reaches the physical limits of its performance is exceptional. In most instances there are also very specific requirements placed on the mechanisms of filtering, transduction and amplification within the receptor cell. In order to make a proper assessment of the performance of the sensor it is necessary to know the biology of the organism as well as the characteristics of the system under investigation. Smart technology in nature is therefore much more complex than in man's technology. The trap is that we will feel we have to copy the biological mechanism slavishly. Or perhaps worse: that we cannot unravel the complexity and miss a simple mechanism which is surrounded by functional irrelevance.

10.2.1.1   *Arthropod Mechano-Receptors*

Arthropod mechano-sensors have been well researched but not much exploited biomimetically. Since they are made from a cuticle which is basically a fibrous composite (insect and spider cuticles are fairly similar; the non-terrestrial crustacea have up to 40% $CaCO_3$) their implementation in

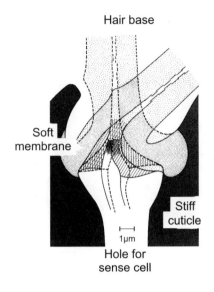

Fig. 10.1   Section through the hair of an insect showing the deformations when it bends [5].

technical systems is relatively simple. The problems are to do with understanding (a) the mechanisms operating in the sensor systems (by no means always apparent) and (b) the degree to which scaling is important. In insects there are three main groups of external sensors: (i) bristles and hairs; (ii) trichobothria; (iii) lyriform organs and campaniform sensilla [3]. Legged robots modelled on insects have been proposed for use on rugged, dangerous, or inaccessible terrain. The agility of insects is probably due mainly to the rich variety of sensory information that is provided by their sense organs. Three classes of mechanosensory organs seem to play an important role in sensory feedback: proprioceptors of movement and mechanical stress and detectors of external contact [4]. These authors point out that the legs of arthropods also provide information about the animal's surrounding — they are part of the general sensory system.

Bristles and hairs (actually not similar to mammalian hairs, but long hollow tapering columns or cones made of cuticle, Fig. 10.1) are commonly set in a socket made of resilin, a rubbery form of insect cuticle which is compliant, has a high degree of resilience and exhibits very little creep.

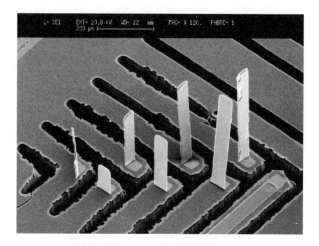

Fig. 10.2   An array of hair-like sensors made from silicon technology [with permission from Dr. Chang Liu, Micro Actuators, Sensors and Systems Group, University of Illinois].

It probably achieves these properties by a combination of controlled cross-linking (covalently with di- and ter-tyrosine [6]) and hydration, which allows the relatively stiff protein chains sufficient steric freedom to exhibit a high degree of rubber elasticity. The hairs when displaced deform a sensory cell within the insect and so generate a signal. When not displaced by external means, the elastic mounting ensures that the hair is in a reference position and so does not need to transmit information about its position. In other words, the structure is so reliable that it generates information only when it is disturbed from its equilibrium position. The hairs can be displaced by external stimuli or by the proximity of other body parts (proprioception) or by air movements. A nanoscale model system has been made (Fig. 10.2) but there are no data about its performance [7].

Trichobothria (Fig. 10.3) are rather unusual types of hair which function entirely as air movement detectors. Adults of the wandering spider *Cupiennius salei* have 936 trichobothria on their legs and pedipalps. The trichobothria are 100–1400µm long and 5–15µm wide (diameter at base). Many of them are bent distally pointing towards the spider's body. Their feathery surface increases drag forces and thus mechanical sensitivity by enlarging the effective hair diameter without increasing mass as much. Typically, trichobothria are arranged in clusters of 2 to 30 which increase in length towards the leg tip. They are tuned to frequency ranges between

40 and 600Hz depending on their length. Because, with increasing length, absolute mechanical sensitivity changes as well, the arrangement of hairs in a cluster provides for a fractionation of both the intensity and frequency range of a stimulus, in addition, in some cases, to that of stimulus direction. The thickness of the boundary layer above the spider leg in oscillating airflow varies between about 2600µm at 10Hz and 600µm at 950Hz, well within the range of hair lengths. Short hairs are as good or better velocity sensors as long hairs but more sensitive acceleration sensors [9]. The spider can detect a buzzing fly with the trichobothrial system at approximately 20cm in all horizontal directions [10]. In sensory terms similar to trichobothria, but larger and an order of magnitude more sensitive, the antennae of male mosquitos are tuned sound detectors. The antennae are covered in long bristles (easily seen with the naked eye) which are stiffly coupled to the antenna rather than articulated as a sensory hair would be. At the base of the antennae is a large bulge (Fig. 10.4) named after its discoverer, Johnston. This bulge contains an array of displacement sensors, collectively known as a chordotonal organ, which transduces the movement of the antenna into nervous impulses [11]. Presumably these various hairs and bristles could also be modelled using standard silica technology, though the resonant properties would be difficult to model since insect cuticle has a specific gravity of no more than 1.5.

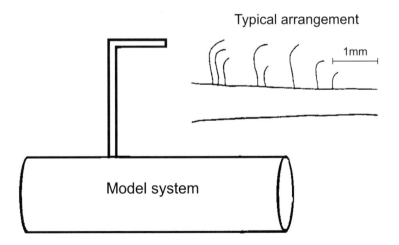

Fig. 10.3   Trichobothria on a spider's leg and a model used to analyse their function [8].

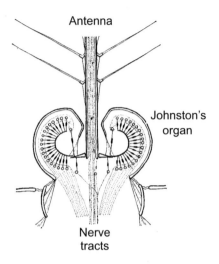

Antenna

Johnston's
organ

Nerve
tracts

Fig. 10.4   Section through Johnston's organ at the base of the antenna of a mosquito [12].

Campaniform sensilla are strain sensors which are basically holes in the
cuticle covered with a bell-shaped (hence campaniform) cap (Fig. 10.5) .
These organs allow the animal to measure displacements in the plane of
the cuticle, and they do it by introducing compliance. The hole which is
the organ actually weakens the surrounding cuticle so that it deforms more
under a given load, thus pushing the dome up and down. The main action
is therefore to rotate the applied deformation through 90° so that in-plane
deformation becomes out-of-plane deformation which can be measured remotely by a cell sitting in the epidermis which produced the cuticle. These
organs are always placed in areas where the load is likely to be the greatest. These are also the places where the structure in which the load is
being measured — the wing or the leg — is most likely to break. Such
an apparent conflict of requirements is a sure sign that there is something
interesting going on. The animal seems to be weakening those areas which
we would expect it most wanted to strengthen. In a simplified model failure
occurs when the strain energy density exceeds a value characteristic of the
material. Thus, the strain energy distributions arising from the presence of
openings can provide a first estimate of the strength reduction they cause.
In a series of finite element models holes were introduced into a composite

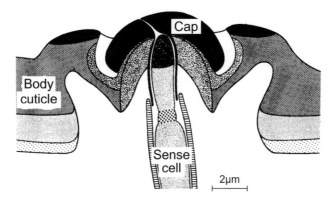

Fig. 10.5   Section through a campaniform sensillum showing the thin area of the cuticle around a hole which has a cap closing it [12].

plate [13]. The holes were considered either drilled (the hole interupted the fibres) or formed (the fibres were deflected around the hole and remained entire) and were circular or elliptical (Figs. 10.6(a) and 10.6(b)). The strain energy is maximum at an angle of 90° at the edge of the opening in all cases, except for a formed elliptical opening parallel to the loading where the maximum is located along the major axis of the ellipse. Drilled holes result in greater strain energy densities near the hole boundaries and steeper gradients of its distribution in comparison with formed holes. In the worst case (elliptical drilled opening orientated normal to the direction of loading) the energy density increases twenty-fold compared with the intact plate. With formed holes the effect is not so pronounced — the strain energy maximum is reduced to about half that found with the drilled hole. The distribution of strain energy is also distributed much more broadly across the specimen when the hole is formed rather than drilled.

There are many ways to measure the deformation of the hole, including the dome system used by insects. Possibly the most obvious is detection of changes in birefringence of plastic in the hole, or change in wavelength of light reflected from an optical grating placed within or around the hole. the hole can therefore be interrogated remotely, which reduces complexity, and therefore costs, greatly.

The concept of the holes can be developed in a number of ways; insects have arrays of holes in major sense organs such as the haltere (flight balancing organ of flies, Fig. 10.7) but the design advantages of doing this have not yet been explored.

In spiders a similar type of organ, the lyriform organ, is composed of a series of slits, each of which is basically the same as a campaniform sensillum but with a much higher aspect ratio. The slits are usually aligned parallel to each other (Fig. 10.8). The mechanical implications of various types of slit arrangements found among the strain-sensitive slit sensilla in the arachnid exoskeleton were studied by measuring the deformation of models made as groups of parallel slits cut into Perspex disks. Static loads were then applied from varying directions. The resulting deformation of the slits was measured in terms of compression and dilatation, respectively and compared to that of a model with only one slit [16]. In simple models with three or five slits of equal length (Fig. 10.9) the deformation of an individual slit is greatly reduced by the presence of neighbouring slits closely arranged in parallel. The peripheral slits deflect more than the intermediate slits. In large groups they are deformed even more than a single, isolated slit. In all cases the maximum deflection of the slit is imposed by a global displacement orientated normal to the long axis of the slit. The slit does not deform uniformly along its length, and the greatest deformation does not always occur half way along the slit. The arrangement of the slits leads

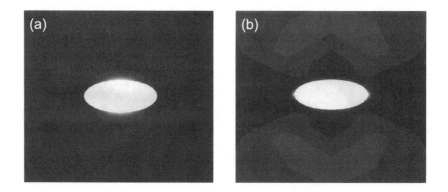

Fig. 10.6 Finite element model of an oval hole (a) cut through a plate of fibrous composite material and (b) moulded into a plate of fibrous composite material so that the fibres are continuous around the hole. The fibres and direction of straining are along the long axis of the oval [14].

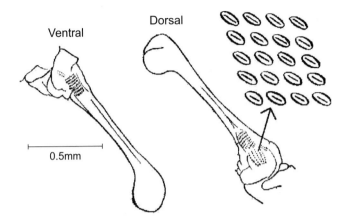

Fig. 10.7 Two views of the haltere (balancing organ — a modified wing) of the blue-bottle fly, *Lucilia*. One of the arrays of campaniform sensilla is shown enlarged [8].

to a much greater range of effective stimuli at a specific site and a marked stimulus intensity fractionation by different slits within the same group. In all slits compression results from a range of load angles larger than 120°. In arrangements with a regular increase in slit length and a triangular outline shape deformability differs greatly among the slits at all load angles. The

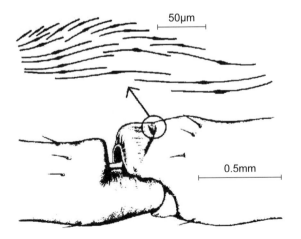

Fig. 10.8 Arrangement of slits in a spider's leg sense organ at the junction between the tarsus and metatarsus [8].

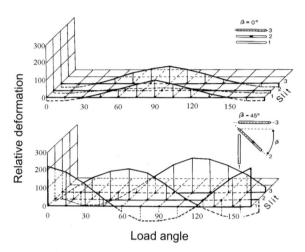

Fig. 10.9   Results from a model system of spider sense organ generated by cutting slits in a sheet of Perspex and applying force at angles varying from parallel with, to perpendicular to, the long axis of the slits [15].

slit configuration with a heart-shaped outline accommodates a large spread of load angles at which the compression of the different slits is highest. These properties suggest different arrangements for the design of different strain measuring problems [15]. These systems have not been analysed further and in particular have not been modelled in an anisotropic plate as has been done with holes of lower aspect ratio.

Buprestid beetles of the genus *Melanophila* (Fig. 10.10) possess paired pit organs next to the base of the middle legs. Each pit contains about 70 tightly packed infrared sensilla that enable the beetles to detect forest fires at long range. A single sensillum consists of a spherule of cuticle about 15µm in diameter. Distal processes of two enveloping cells surround the entire spherule in the form of a flattened protoplasmatic layer with the exception of a small apical stalk connecting the spherule to the outer cuticle. The spherule is innervated by a single sensory neuron. The sensilla respond (measured from the nervous supply to the organ) to infrared radiation and can respond to a stimulus of only 2ms duration. The sensilla have probably evolved from mechanoreceptors so that mechanical events are still part of the transduction process. They strongly resemble the basal regions of hair

Fig. 10.10 *Melanophila acuminata*, the fire beetle. The sense organs are on the under-side of the animal.

mechanoreceptors (sensilla trichodea) in their immediate neighbourhood and so are probably derived from them [13; 17]. Their mode of functioning is still unclear, though since they are being studied intensively, this will not remain so for long.

10.2.1.2 *Vertebrate Sensors*

There are many types of sensors amongst the vertebrates, some of which are being copied or used for inspiration.

Electrosensing

Teleost (bony) fish, elasmobranchs (sharks and rays) and the duckbilled platypus (and probably many more types of animal) have an electric sense. It is best developed in the elasmobranchs, which have rows of pit organs (ampullae of Lorenzini) which can detect electric fields as weak as 5nVcm^{-1}, and so detect the fields induced through their bodies as they swim through the Earth's magnetic field. They can use this sense to detect the presence of prey and there is evidence that they use it in navigation. However the electroreceptors can not measure DC voltages so that a voltage due to water flow in the ocean is not uniquely interpretable in terms of the speed and direction of flow at the point where the electrical measurement is made.

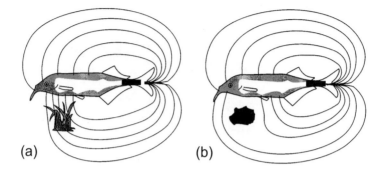

(a)   (b)

Fig. 10.11   Electric fields of a *Gnathonemus petersii* distorted by (a) a water plant (good conductor) and (b) a stone (bad conductor). The fish is viewed from the side with the electric field lines drawn as thin lines and the receptive parts of the body surface shown in grey [20].

Perhaps the cue is the directional asymmetry of the change in induced electroreceptor voltage during turns. A neural network could use this cue to determine swimming direction by comparing electrosensory signals and signals from the semicircular canals, which function as an accelerometer [18]. Weakly electric fish use active electrolocation — the generation and detection of electric currents — to explore their surroundings (Fig. 10.11). Although electrosensory systems include some of the most extensively understood circuits in the vertebrate central nervous system, relatively little is known quantitatively about how fish electrolocate objects. A combination of techniques, including field measurements, numerical and semi-analytical simulations, and video imaging of behaviours is leading to the identification of image features and computational algorithms that could reliably encode electrosensory information [19].

Thermal Imaging

Boid snakes possess unique infrared imaging pit organs. The surface of these organs scatters or reflects electromagnetic radiation of specific wavelengths. The skin of the pit organ of these snakes is covered with arrays of pore-like structures called micropits. These micropits average 319nm in diameter and 46nm in depth and are spaced about 808nm from each other. These arrays seem to function as spectral filters or anti-reflective coatings for incident electromagnetic radiation and so may enhance their absorption of infrared radiation, thus increasing their sensitivity as heat receptors [21]. Apart from this specialisation, their sensitivity as heat receptors is not outstanding, except that there are more of them in a small area than in, say, a patch of sensitive human skin, and they are considerably nearer the surface of the skin.

10.2.2   *Integration and Coding*

The assumption here is that evolution will favour nervous systems in which information is transmitted at a high or optimum rate with ample redundancy. There is little point in reviewing the mechanisms of nervous transmission, nor of central processing. These are either readily available or irrelevant. Is the information transmitted by an ensemble of neurons determined solely by the number of pulses from each cell, or do correlations in the emission of pulses also play a significant role? A crucial point is how the information conveyed by individual cells combines together. If many cells in the sample carry similar information, then the code is redundant, and there is not much more information in the population than that present in a single cell. Where the information from different cells is independent the information increases linearly with the number of cells in the population. If some information is available only by looking simultaneously at the responses of different neurons with, for example, some of the information being available only from the relative timing of pulses from different cells, the code is said to be synergistic, and more information is available from the population than one would obtain by the sum of the information obtained from each neuron alone. Over short time-scales correlations cannot dominate information representation, but stimulus-independent correlations may lead to synergy. But only certain combinations of the different sources of correlation result in significant synergy rather than in redundancy or in negligible effects [18].

10.2.3   *Actuation*

10.2.3.1   *Skin*

The reorientation of the main fibrous component, collagen, in skin when it is stretched can be greatly affected by the introduction of a short split or

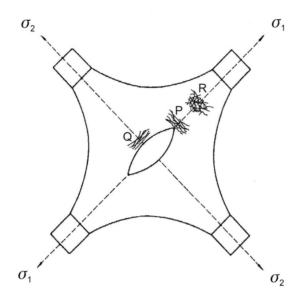

Fig. 10.12   Orientation of collagen fibres around a slit in a sheet of aortic media from a
pig [22].

notch [22]. The notch induces a zone of concentrated stress at its tip which
decreases very quickly with increasing distance ahead of the tip. The effect of this stress (detected as reorientated collagen fibres) disappears only
0.5mm ahead of a 5mm crack in aortic media (a collagenous layer in the
aorta) of the pig which was stretched biaxially by about a third of its original length (Fig. 10.12). The stress concentration has its maximum local
stresses perpendicular to the long axis of the crack, so that the collagen fibres align in a direction perpendicular to the crack tip (point P). If the crack
is to advance it now has to cross many more fibres and will require much
more energy to do so. With a random arrangement of fibres (as at point R),
the material is therefore initially isotropic so that initiation and propagation of the crack will be equally difficult in all directions. At point Q the
collagen is orientated parallel to the long axis of the crack. This trick places
skin firmly in the category of a responsive material. The important characteristics of skins leading to this type of property are heterogeneity (stiff
fibres in a softer matrix) and a degree of mobility of the fibres. Depending
on the resistance to crack propagation and degree of initial orientation of

the fibres, skins can be stretched to high (10 to 50%) strains which are associated with an elastic modulus increasing with deformation, giving a J-shaped stress–strain curve. This curve has many interesting characteristics, and seems independently to be associated with high toughness [23], perhaps an underlying reason for using such an arrangement of fibres in an outer covering.

10.2.3.2 *Deployable Structures*

"Intelligence" in deployable structures will reside in their ability to respond to local conditions. Concepts based on the folding and hinging mechanisms of insect wings would be very interesting since the wing is actuated only from the base so all shape changes which occur in the main part of the wing must be due to the interactions of folding and elastic mechanisms. The wing is hinged at three independent points at the base, so the main lamina can be modelled as three areas, mutually hinged along their edges, capable of changing their spatial and phase relationships with each other during the stroke cycle. Distally folds which cross the two primary folds, which delimit these three areas, can be activated or suppressed by the action of the primary folds [24]. Add elastic mechanisms to this ability [1; 25] and there appears a flying surface which, because of its controlled deformability, can respond instantly to local conditions partly innately due to its aeroelastic response during the wing stroke and partly under remote control from the muscles at the wing base. Some of the shape-locking mechanisms, based on control of elastic buckling, may also prove interesting [25]. They remain to be analysed.

A concept which does not seem to have been explored, which occurs more frequently than one might think in nature, is using hydraulic pressure to store strain energy in an elastic component. This is the underlying principle of the Venus Fly Trap (Fig. 10.13) and very probably in other micromechanisms involved with pollination, for instance in orchids where a mechanism (the pollinium) in the pollen-bearing part of the flower sticks on to the visiting insect, then bends over (Fig. 10.14) so that it can stick pollen on to the female part of the next flower which the insect visits [26]. The elastic energy store is the cellulose in the walls of the cells containing the pressurised liquid; the liquid is more or less incompressible. This approach has the advantage of power amplification, so that the strain energy can be accumulated at a low work rate and released suddenly.

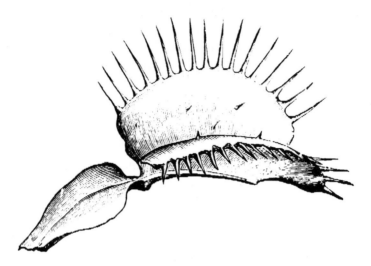

Fig. 10.13　Open leaf of a Venus Fly Trap.

10.2.4　*Implementation*

Developing the concept that the simplest form of intelligent biomimetic system is likely to be the most successful, what systems are there in nature which offer the simplicity and functionality of the bimetallic strip? Self-assembling materials and structures, and in particular liquid crystal structures, offer the most advantages not only because we can make and manipulate them with great variety [27] but because they are ubiquitous in nature and may well be found at the root of most structural elements. If one takes this further and investigates the possibilities inherent in the phase separation and morphological driving forces in block copolymer systems [28], then nanocomposites are relatively easy to make. Once one has an array of separate elements, then extra information is available as the structure of the array interacts with external stimuli (light, mechanical loads, etc.).

10.2.4.1　*Liquid Crystals*

Liquid crystals are now seen to be almost everywhere you care to look in biological systems. Liquid crystals "self-assemble" from a disordered state, and so represent a way in which order, and therefore morphology, can be

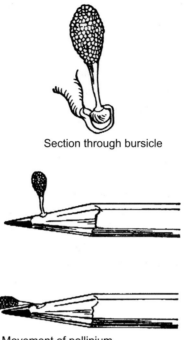

Section through bursicle

Movement of pollinium

Fig. 10.14   Pollinia [26].

generated in a purely chemical system. Since "living" tissues are made of chemicals, and "life" is achieved as a result of the ordering and partitioning of these chemicals, there is much interest in any mechanism for achieving that order. And liquid crystals can generate a variety of types of order from relatively simple molecules, and can transform from one type of order to another (Fig. 10.15) in response to changes in external conditions (e.g. changes in salt concentration, temperature, pH). Liquid crystals conform to one of the criteria of biological systems — that what they do should be achieved with the minimum expenditure of energy. For instance the energy required to convert a nematic liquid crystal (in which rod-like molecules are held parallel to each other) into a helicoidal conformation (where the rods are arranged in layers at successive angles to each other) with a pitch of

1µm is 10^5 times less than the amount of energy needed to induce nematic order in an initially random system. And the generation of the nematic system can be rendered far more energy-efficient by orientating the molecules against a flat surface. Self-assembly systems for the generation of biological materials are more energy efficient than those which cannot generate their own order and therefore need enzymatic control and energy input. On the face of it, therefore, liquid crystalline structures should be ubiquitous. They offer advantages at the morphological and energetic levels. The problem remains that the mechanisms by which structures are generated in biological systems are not clear. But it may be that we have to think that liquid crystalline structures are low energy in terms of structural maintenance rather than generation, so that the cell drives a structure towards a liquid crystalline morphology but stability comes from the intrinsic properties of that morphology. The output from a liquid crystal system, in technical terms, can be light (e.g. changes in form birefringence, colour), dielectric properties, mechanical properties (rheological or stiffness), etc.

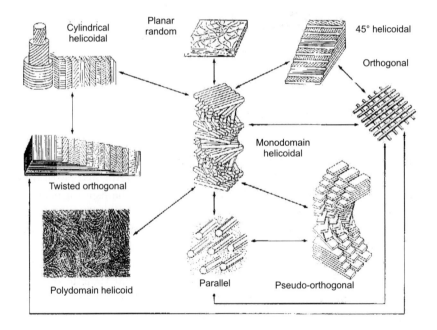

Fig. 10.15   Transformations of liquid crystal structures [29].

The similarities between liquid crystals and a biological material — insect cuticle — were first noticed by Charles Neville (a zoologist) and Conmar Robinson (a polypeptide chemist) in 1967 [29]. The optical properties are very similar; in insect cuticle the parallel and helicoidal fibrous structures rotate the plane of polarisation of light in the same way that nematic and cholesteric liquid crystals do. The difficulty in this comparison is that the conformation of liquid crystals is controlled from the molecular level, whereas the orientations in insect cuticle are apparently controlled at the just-sub-micrometer level — a difference of at least two orders of magnitude. In some cuticles with reversible stiffness, the insect can increase the water content so that the modulus decreases. This happens in the blood-sucking bug *Rhodnius*, for instance, which can change the pH of the cuticle from about 7 to below 6, thereby increasing cuticular water content from about 26% to 31%, dropping its stiffness from 250MPa to 10MPa and increasing its extensibility from 10% to more than 100% [30; 31]. It is able to do this because the proteins of the cuticle are hydrophobic and so can have their water-binding characteristics changed very readily [32].

The intelligent controlled use of plasticisers in composites to mediate shape change has not been developed. There are times when this ease of transformation is an advantage, such as in the development of dogfish egg case [33] and the production of silk [34]. But there are equally other times when the resulting structure has to be stable so that it can carry or generate forces, and under those circumstances the order has to be locked in to the structure by processes which lead to cross-linking of the components [35].

The morphology of liquid crystals can be modified by changes in concentration, temperature, pH and salinity. Since the regular packing of a liquid crystal represents a higher density and a lower energy state, higher pressure and lower temperature will favour more liquid crystalline structure. The protein of the mantis egg case is helicoidal above pH 5, isotropic below. Collagen is a liquid crystal as is amply shown in the dogfish egg case [34] and changes its form of packing and hydration with changes in concentration and type of external salt. So there are many ways of affecting the packing patterns, which in turn can be used to indicate the surrounding conditions when they formed their morphology. These changes can therefore be used to transduce information about their surroundings, and become the initial stage in a sensor. The interrogation of the sensor is easily made with polarised light, which is non-contact and can be remote.

10.3   Conclusions

Natural systems are the paradigm both for smart systems and for robotics. It is proper therefore to turn to nature to see what ideas can be extracted for use in these mechanisms. Nature can teach not only about materials and structures, but also about methods. It can teach about the production and manipulation of such materials in an intelligent manner. Natural systems are self-organising at a number of levels. At the molecular and microstructural level this is very probably controlled or influenced by liquid crystal materials and structures. But at the macrostructural level it is almost certainly controlled by the emergent properties of a population of individual constructional effectors. This notion may be more difficult for an engineer (who needs to know that the structure will carry the design loads reliably) and probably even more so for the user (who has to be confident in the structure and know whom to sue when it fails!). Probably the self-designing structure will emerge in an aerospace environment where payload costs, the lack of gravitational forces and the low maintenance requirement will combine to make such an option very attractive. If we can do it. One problem remains. How does one set about scrapping and recycling a self-sensing and self-repairing structure?

Bibliography

[1]   Haas, F., Gorb, S. and Blickhan, R. (2000) "The Function of Resilin in Beetle Wings, *Proc. Royal Soc. London B*, **267**, 1375–1381.

[2]   Bialek, W. (1987) "Physical Limits to Sensation and Perception", *Ann. Rev. Biophys. & Biophysical Chem.*, **16**, 455–78.

[3]   Keil, T. A. (1997) "Functional Morphology of Insect Mechanoreceptors", *Microscopy Res. & Technique*, **39**, 506–531.

[4]   Delcomyn, F., Nelson, M. E. and Cocatre–Zilgien, J. H. (1996) "Sense Organs of Insect Legs and the Selection of Sensors for Agile Walking Robots", *Int. J. Robotics Res.*, **15**, 113–127.

[5]   Thurme, U. (1965) "An Insect Mechanoreceptor Part I: Fines Structure and Adequate Stimulus", *Cold Spring Harbor Symp. on Quantitative Biology*, pp. 75–82.

[6]   Andersen, S. O. and Weis–Fogh, T. (1964) "Resilin, a Rubber-Like Protein in Arthropod Cuticle", *Adv. Insect Physiol.*, **2**, 1–65.

[7]   Goldin, D. S., Venneri, S. L. and Noor, A. K. (2000) "The Great Out of the Small", *Mech. Eng.*, Nov. 2000, 71–79.

[8]   Barth, F. G. and Geethabali (1982) "Spider Vibration Receptors: Threshold Curves of Individual Slits in the Metatarsal Lyriform Organ", *J. Comparative Physiol.*, **148**, 175–185.

[9]   Barth, F. G., Wastl, U., Humphrey, J. A. C. and Devarakonda, R. (1993) "Dynamics of Arthropod Filiform Hairs. II. Mechanical Properties of Spider Trichobothria (*Cupiennius Salei* Keys.)", *Phil. Trans. Royal Soc. London B*, **340**, 445–461.

[10] Barth, F. G., Humphrey, J. A. C., Wastl, U., Halbritter, J. and Brittinger, W. (1995) "Dynamics of Arthropod Filiform Hairs. III. Flow Patterns Related to Air Movement Detection in a Spider (*Cupiennius Salei* Keys.)", *Phil. Trans. Royal Soc. London B*, **347**, 397–412.

[11] Goepfert, M. C., Briegel, H. and Robert, D. (1999) "Mosquito Hearing: Sound-Induced Antennal Vibrations in Male and Female *Aedes Aegypti*",

J. Exp. Biol., **202**, 2727–2738.

[12] Wigglesworth, V. B. (1965) *The Principles of Insect Physiology*, 6th edition, Methuen.

[13] Skordos, A., Chan, P. H., Vincent, J. F. V. and Jeronimidis, G. "A Novel Strain Sensor Based on the Campaniform Sensillum of Insects", *Phil. Trans. Royal Soc. London A* (to appear).

[14] Schmitz, H. and Bleckmann, H. (1997) "Fine Structure and Physiology of the Infrared Receptor of Beetles of the Genus *Melanophila* (Coleoptera, Buprestidae)", *Int. J. Insect Morphol. & Embryol.*, **26**, 205–215.

[15] Barth, F. G., Ficker, E. and Federle, H. –E. (1984) "Model Studies on the Mechanical Significance of Grouping in Compound Spider Slit Sensilla (Chelicerata, Araneida)", *Zoomorphol.*, **104**, 204–215.

[16] Barth, F. G. and Pickelmann, P. (1975) "Lyriform Slit Sense Organs Modelling an Arthropod Mechanoreceptor", *J. Comparative Physiol.*, **103**, 39–54.

[17] Vondran, T., Apel, K. H. and Schmitz, H. (1995) "The Infrared Receptor of *Melanophila Acuminata* Degeer (Coleoptera, Buprestidae) — Ultrastructural Study of a Unique Insect Thermoreceptor and its Possible Descent from a Hair Mechanoreceptor", *Tissue & Cell*, **27**, 645–658.

[18] Paulin, M. G. (1995) "Electroreception and the Compass Sense of Sharks"'", *J. Theor. Biol.*, **174**, 325–339.

[19] Assad, C., Rasnow, B. and Stoddard, P. K. (1999) "Electric Organ Discharges and Electric Images During Electrolocation", *J. Exp. Biol.*, **202**, 1185–1193.

[20] von der Emde, G. (1999) "Active Electrolocation of Objects in Weakly Electric Fish", *J. Exp. Biol.*, **202**, 1205–1215.

[21] Campbell, A. L., Bunning, T. J., Stone, M. O., Church, D. and Grace, M. S. (1999) "Surface Ultrastructure of Pit Organ, Spectacle, and Non-Pit Organ Epidermis of Infrared Imaging Boid Snakes: A Scanning Probe and Scanning Electron Microscopy Study", *J. Struct. Biol.*, **126**, 105–120.

[22] Purslow, P. P., Bigi, A., Ripamonte, A. and Roveri, N. (1984) "Collagen Fibre Reorientation Around a Crack in Biaxially Stretched Aortic Media", *Int. J. Biological Macromolecules*, **6**, 21–25.

[23] Mai, Y. -M. and Atkins, A. G. (1989) "Further Comments on J-Shaped Stress–Strain Curves and the Crack Resistance of Biological Materials"'", *J. Phys. D*, **22**, 48–54.

[24] Wootton, R. J. (1981) "Support and Deformability in Insect Wings", *J. Zoological Soc. London*, **193**, 447–468.

[25] Haas, F. (1994) *Geometry and Mechanics of Hind-Wing Folding in Dermaptera and Coleoptera*, MSc thesis, University of Exeter, Exeter, UK.

[26] Darwin, C. (1877) *The Various Contrivances by which Orchids are Fertilised by Insects*, John Murray.

[27] Donald, A. M. and Windle, A. H. (1992) *Liquid Crystalline Polymers*, Cambridge University Press.

[28]   Arridge, R. G. C. and Folkes, M. J. (1972) "The Mechanical Properties of a 'Single Crystal' of SBS Copolymer — A Novel Composite Material", *J. Phys. D*, **5**, 344–358.

[29]   Neville, A. C. (1993) *Biology of Fibrous Composites: Development Beyond the Cell Membrane*, Cambridge University Press.

[30]   Reynolds, S. E. (1975) "The Mechanical Properties of the Abdominal Cuticle of *Rhodnius* Larvae", *J. Exp. Biol.*, **62**, 69–80.

[31]   Hillerton, J. E. (1979) "Changes in the Mechanical Properties of the Extensible Cuticle of *Rhodnius* through the Fifth Larval Instar", *J. Insect Physiol.*, **25**, 73–77.

[32]   Urry, D. W. (1995) "Elastic Biomolecular Machines", *Scientific American*, Jan. 1995, 44–49.

[33]   Knight, D. P., Feng, D., Stewart, M. and King, E. (1993) "Changes in Macromolecular Organization in Collagen Assemblies During Secretion in the Nidamental Gland and Formation of the Egg Capsule Wall in the Dogfish *Scyliorhinus Canicida*", *Phil. Trans. Royal Soc. London B*, **341**, 419–436.

[34]   Knight, D. P. and Vollrath, F. (1999) "Liquid Crystals and Flow Elongation in a Spider's Silk Production Line", *Proc. Royal Soc. London B*, **266**, 519–523.

[35]   Neville, A. C. (1975) *Biology of Arthropod Cuticle*, Springer.